The Chief and I

To Bob and Adelle,
my neighbors on
The Mattaponi!

Karen Tootelian

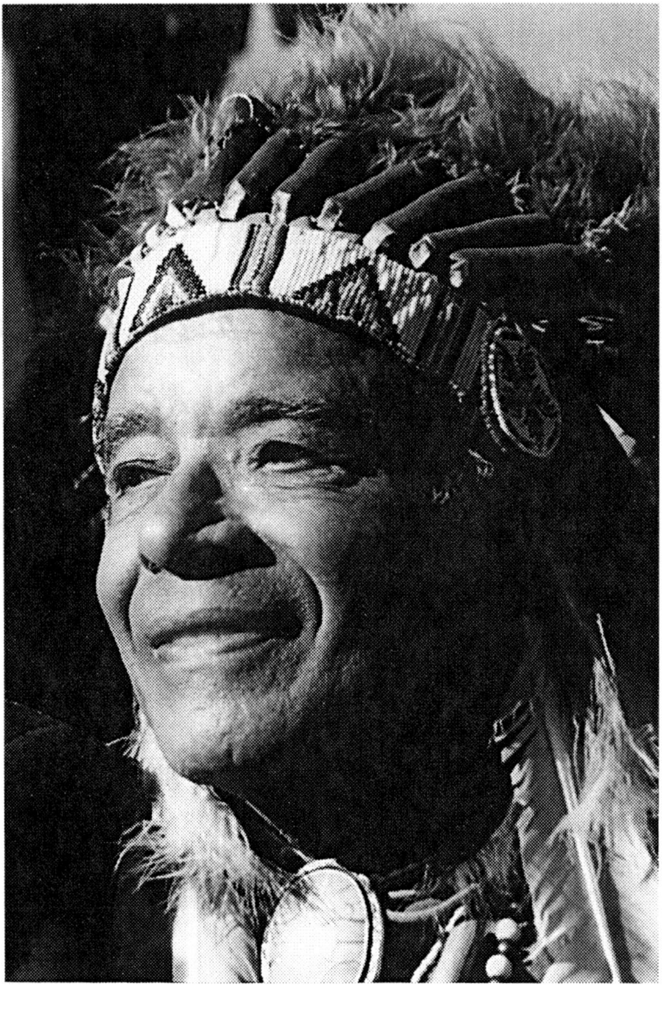

The Chief and I

By
Karen Tootelian

Brandylane Publishers, Inc.
Richmond, Virginia

Copyright 2007 by Karen Tootelian. All rights reserved. No portion of this work may be reproduced in any form without the written permission of the publisher.

ISBN 1-883911-75-3

Library of Congress Control Number
2007927024

Front cover watercolor: "Mattaponi Sunset" by Anne Careatti
Back cover photo of Webster Little Eagle:
www.baylink.org/Mattaponi
Back cover author photo by Warren Mountcastle

Brandylane Publishers, Inc.
Richmond, Virginia
www.brandylanepublishers.com

Dedicated to

Webster Little Eagle Custalow
November 14, 1912—March 21, 2003
~ Chief of the Mattaponi Indian Tribe from 1977
through 2003 ~

and his wife

Mary White Feather Adams Custalow
September 1, 1912—March 31, 1993

HOUSE JOINT RESOLUTION NO. 22
Commending Chief Little Eagle Webster Custalow of the Mattaponi tribe. Agreed to by the House of Delegates, January 21, 1994 Agreed to by the Senate, January 27, 1994

WHEREAS, 1994 will mark Chief Little Eagle Webster Custalow's forty-fifth year as a leader of the Mattaponi tribe; and

WHEREAS, Webster Custalow was born on the Mattaponi Indian reservation in King William County on November 14, 1912, the son of Chief Haskanawanaha George Forest Custalow and a descendant of the Mahayough (Major) family chieftain lineage of the Mattaponi tribe which is traced back to 1658; and

WHEREAS, Webster Custalow completed with distinction the six grades offered at the time by the one-room reservation school and soon embarked on a career as a logger and timberman renowned for his ability to cruise timber and estimate pulpwood; and

WHEREAS, Little Eagle Webster Custalow became assistant chief of the Mattaponi tribe in 1949 and was elected chief in 1977; and

WHEREAS, Chief Little Eagle Webster Custalow

has worked unstintingly to improve the health and living standards of the Mattaponi reservation, implementing a change from shallow to deep wells to provide purer domestic water supplies, spearheading efforts to relocate the King William County landfill that was contaminating the reservation's water supply, supervising the rehabilitation and replacement of older housing on the reservation, restoring the tribal community building to accommodate pottery making and other craft work, and building a fish hatchery on the reservation to spawn and hatch shad; and

WHEREAS, Chief Little Eagle Webster Custalow for a half-century has worked to improve relations between the Mattaponi reservation and King William County and also has made numerous civic contributions to the wider community, such as serving on the board to develop the King William Community Health Service and helping to start the West Point Parent Child Development Center; and

WHEREAS, Chief Little Eagle Webster Custalow has dedicated himself to improving the environment, working to restore the osprey and bald eagle to the King William County area and the shad supply to Virginia's rivers and to preserve the ecology of the Mattaponi River generally; and

WHEREAS, Chief Little Eagle Webster Custalow is well-known as a representative of the Mattaponi in particular and Virginia's Native Americans in general in educational and civic forums and has eagerly continued the traditions that maintain the treaty relationship established between the Commonwealth and the Mattaponi in 1677; and

WHEREAS, Chief Little Eagle Webster Custalow was saddened this past year by the death of his wife of 62 years and mother of their nine children, Princess White Feather Mary L. Custalow, on March 31, 1993; and

WHEREAS, at the age of 81, Chief Little Eagle Webster Custalow continues his service as a deacon of the Baptist church at the Mattaponi reservation, his active leadership on the reservation, and his participation in the wider community; now, therefore, be it

RESOLVED by the House of Delegates, the Senate concurring, That the General Assembly commend Chief Little Eagle Webster Custalow for the outstanding leadership he has provided the Mattaponi tribe, congratulate him for a distinguished career of public and community service, and wish him many more years of success; and, be it

RESOLVED FURTHER, That the Clerk of the House of Delegates prepare a copy of this resolution for presentation to Chief Little Eagle Webster Custalow as an expression of the deep admiration and affection in which he is held by the members of the General Assembly.

Table of Contents

Foreword	xiii
2002 ~ The Chief and I	1
2003 ~ This Wild Place	41
2004 ~ Mysteries	77
2005 ~ Riversong	109
2006 ~ Blooming	139
Appendix	146
Bibliography	152

Foreword

Chief Little Eagle was a small person in stature but a giant in the public's eye. Because of his powerful, yet humble presence, he was loved and respected by all. He was truly a people person and loved to talk and give of himself.

As he grew older and weaker, he became fearful of hospitals and doctors, as he always said he wanted to pass from this life in his home. He was uneasy when nurses would come to see him, fearful they might say he should go back to the hospital. He resisted the family securing outside help for him. Many came, and many went because he didn't want to feel he was losing his independence. He didn't feel comfortable with strangers telling him what he could and could not do in his own home.

I remember the first day I took Karen in to introduce her and let him know she would be helping care for him. I figured as in the past, he would have a negative attitude and say he didn't need anyone to help. He usually made his point in a kind and gentle but firm way. I went into the living room and let them talk. The more they talked and laughed, I could feel there was an instant mutual bond, trust, and understanding between them. He did not resist her helping him.

As time went on, I could see that Karen understood him and tried to give him what he wanted and not what she thought he needed. She truly cared for him as a person and elder, and he could sense that. He enjoyed having her there and appreciated her kindness. Because she was not there everyday, he would ask me, "When is the black-haired girl coming again?" When I told him, his eyes would brighten up and he would say, "She's a nice person."

I sensed that during their visits, they both learned something from each other, and both understood each other.

—Chief Carl Lone Eagle Custalow

Acknowledgments

For my husband, Lee Westermann's, loving encouragement and support, I am deeply grateful,

and for the dear friends who insisted that I share my words.

2002 ~ The Chief and I

In the summer of 2002, life presented me with a rare gift. I began a friendship with the eighty-nine-year-old Chief of the Mattaponi Tribe. As I had done for many years, I kept a journal—this one about the time the Chief and I spent together.

January 1, 2002 *Shooting star*
I saw the most wonderful shooting star tonight. I glanced outside the bedroom window and saw a light flashing through the sky. I watched, and the tail exploded and broke apart. Happy New Year.

March 7, 2002 *Touching everything*
I love being still and quiet and watching, warm sun on skin, the river. God, it is so incredible, the simplicity, the layers of the earth. It moves inside of me like breathing and blood. When I am alone at this river, I sit on the ground and touch everything—the small new plants, autumn's dead leaves, spirals of running cedar, the bark of trees. I smell the cool scent of winter leaving the earth.

April 13, 2002 *Mattaponi River*
What were you a hundred years ago? Two hundred, four? Who walked along your banks, at the water's edge? Who swam and laughed in your waters? I come to sit in the forest, to look out from this high bank, to listen, to lose myself from the world. Yearning for a world untouched by man's need to order, to control, to erase the rhythms of nature. Geese are calling.

August 20, 2002 *My first day with Little Eagle*
My first day with Chief Webster Little Eagle Custalow of the Mattaponi Tribe, born November 14, 1912. I am staying with him a few hours on Tuesdays and Thursdays. Staying with the Chief came about by chance. I am friends with his son, Carl Lone Eagle, and he spoke several times of needing someone he could trust to help with his father. I offered, and Carl accepted. He and his father both live on the Mattaponi Reservation, and Carl is the primary caretaker for his dad. He is also the acting Chief, and this entails tremendous responsibility. Carl and his son Todd have been fighting the Newport News reservoir proposal since 1996, and that takes an enormous amount of time and energy. It was Chief Webster who initiated this phase of the battle in 1996, some say breaking the Tribe's centuries of silence against political actions affecting them.

The Chief and I

I stopped at Carl's house first, and he said that he had told his dad the night before that someone was coming to help with his care. His father did not want anyone. Carl was warning me, and he seemed a little worried. We both drove to the Chief's house, and Mr. Custalow came from his bedroom, making his way carefully with his cane. I was happy to see him and hoped our first meeting would go well. We all sat at the kitchen table. He looked at me, was polite but quiet. Then he gave me a sweet, shy smile. I liked his eyes. He was clean shaved—a task the Chief likes Carl to do. We talked in snippets about his two dogs, Gomer and Queenie, and about his medicine. He devotes himself to his dogs. The older Queenie is a cocker and a little high strung. Gomer looks to be part cocker mixed with something else, and younger. He is a sweet, friendly dog with shiny, black eyes.

Carl stayed. I asked the Chief what he had done for a living. He told me he had had a logging business, how back then, there was no fancy equipment to move the logs. The labor had been hard and much more dangerous. He went on to talk about the German workers who had once been prisoners during WWII. As he became more involved in talking, Carl said he would be going. The Chief continued and spoke highly of the German workers. He cried about how dear they were, how he cared for them so much, and they cared about him. He said that if you treat people kindly, they will do the same to you, and these men were grateful to be treated so generously. He said they loved Marlboro cigarettes and chocolate, and he gave them these things. He said they would do anything for him.

The Chief's daughter, Shirley Little Dove, came over. She is the oral historian, designated as such by her grandfather when she was four years old. She travels to schools in Virginia and to Williamsburg and Jamestown teaching about the Mattaponi history and their traditional way of life—"We are the people of the river," is the message Shirley Little Dove speaks. She has the most beautiful blue-green eyes. I met her about twenty-five years ago, and I was struck then by her beauty, and she is still a beauty. We talked, and she spoke softly. She asked me if I sang, and I said yes. She liked that and was hoping I would sing while I was there with her dad. I had thought of that, too. She left, and I fixed some lunch.

The Chief told me that Shirley worked really hard traveling around Virginia and caring for her grandchildren at times. The Chief talked of his love for Carl, their strong connection, and how Carl was a part of him. He teared up again. Carl had told me the same thing. Caring for his aging father was not always easy, but the bond and commitment were of the utmost importance and a matter Carl met on his own terms. I understood this clearly.

I was touched by the Chief's gentleness and kindness to me. He told me he was there to make me happy. He told me "to eat darlin," to take whatever I wanted in the house. When we met Carl last summer, he gave us fresh vegetables from his garden and fish he had caught, a whole cooler full. He displayed an unusual generosity all along, and when I mentioned it to him, he said that was their way. The Chief talked about how blessed his life had been, and my thought was of all that had been taken from them.

When I left, Mr. Custalow said he had enjoyed having me, and later on, Carl told me that his dad spoke highly of me and so had his sister.

First impressions: love, caring, humor, kindness.

It was our belief that the love of possessions is a weakness to be overcome. Its appeal is the material part, and if allowed its way, it will in time disturb one's spiritual balance. Therefore, children must early learn the beauty of generosity. They are taught to give what they prize most, that they may taste the happiness of giving. If a child is inclined to be grasping, or to cling to any of his or her little possessions, legends are related about the contempt and disgrace falling upon the ungenerous and mean person.... The Indians in their simplicity literally

give away all they have—to relatives, to guests of other tribes, or clans, but above all to the poor and the aged, from whom they can hope for no return.
    —Ohiyesa, Santee Sioux

August 29, 2002 *Strong hands*
Mr. Custalow's. Strong hands, beautiful skin. I can only imagine what his hands have accomplished. How he loves his dogs and they love him; they are always at his side. He feeds them bits of whatever he is eating. They lie down with him when he sleeps—Gomer on the bed with her head on his leg, Queenie beneath the bed.

What is man without the beasts? If all the beasts were gone, men would die from great loneliness of spirit, for whatever happens to beasts also happens to man. All things are connected.
    —Chief Seattle, Suqwamish and Duwamish

September 2, 2002 *Just curious about things*
Mr. Custalow was so happy to see me yesterday. I went for a visit. I was so happy to see him too. There is surely a wonderful spirit to this man, and we seem to have connected right away. We hugged and kissed each other on the cheek. I took him two leaves, red and yellow, because "you are curious" I said. He had

told me, "You know, I am just curious about things," in a voice full of wonder and purity. He has a fine vocabulary. And it is not just the vocabulary but how he uses the language, his cadence, his wisdom, and experiences all woven together. Carl says he was a great storyteller and people loved to hear him tell stories. Recently, a mother dolphin and her baby were spotted in the river not far from the reservation. The mother was later discovered dead upriver with big shark bites in her stomach. It is thought that the bite occurred before coming up into the river, and she died in the Mattaponi. No one knows anything more about the baby. I asked the Chief about the dolphin. He said he thought the "adverse" weather caused the dolphin to come as far up the river as it had. This is a summer of drought.

The Chief was telling me too that Carl bathes and shaves him and "spruces" him up and he likes that. He gets a twinkle in his eyes and a wonderful grin. I told him I would wash his hair on Tuesday to "spruce" him up, and he laughed.

September 5, 2002 *Learning together*
Sleeping when I arrived. I dusted until he woke. He had a hard time waking, seemed groggy. I rubbed his back and arms and legs to get his circulation going. We went into the living room for a while,

and I convinced him to go for a drive. He coaxed his "pups" into his bedroom, and we closed the door and made our great escape down the front steps where there is a rail. Apparently, a couple of years ago he fell and broke his neck and had a stroke or a series of small strokes. These things slowed him up quite a bit. Before that, he was still driving his black Toyota pickup (which he loves) and riding his lawn mower. He was shad fishing until he was eighty-five years old. That is amazing to me. From what I have seen, shad fishing is true hard work, with little sleep, that begins in March and goes through May.

He loves going for a drive but does not want to leave his dogs, as his first priority most of the time is his dogs. But we got in the car and drove to Carl's to say hello, then drove on. We stopped to talk to a neighbor. The Chief had a pleased expression. We drove slowly to Wakema (an old house near the reservation). Pleasant weather, sunny and warm, not at all hot. We talked to the caretaker who knew Mr. Custalow; we admired the soybeans and the river. Then we went back to the house. I fixed him stew, bread and butter, fruit, egg, and cupcake. He likes bread and butter with each meal.

He let me wash his hair in the kitchen, and I figured out how to take his shirt off so as not to hurt his shoulder—I am learning. We are learning together about life's pace. He said he loved having me wash his hair. I massage his scalp, then rub ointment into his hurt shoulder. We sat in the den, and he talked but was so tired. I lifted his legs and rubbed them, held his hand. He talked to me about how he loved to see me, my dark hair, to hear me, and he knew things were okay then. I helped him to bed, and he told me how he loved me and I meant so much to him. I told him I loved him too and kissed his cheek. He kissed mine. I cooked him some apples for dinnertime and left for the day. There, life is simple. We talk. I listen. There is touch. I prepare good foods for him. When he talks, and some of it makes sense and some of it is unclear to me, I try to hear his heart. This is life. It is true. When I leave there, I am calm. I am open. I can write.

My heart laughs with joy because I am in your presence. Ah, how much more beautiful is the sun today. . .
—Chitmachas Chief

September 9, 2002 *A smile that lights my heart*
Visited Mr. Custalow. Carl and his younger son Chad (fourteen) were there. Mr. Custalow was so happy to see me; gave me a big hug and said he had been waiting for me. I sat next to him and held his hand and made him laugh. I took him two beautiful leaves, red and yellow, "Because you told me you are

curious," I told him again. He smiled his wonderful smile. It truly lights his face. And my heart.

September 10, 2002 *"It was long ago, darlin'."*
Busy Chief day. I baked chereg—an Armenian braided bread, and his house smelled like sweet bread. While dusting his figure of an Indian on a horse, I asked if he had ridden horses. He perked right up and said yes, he had ridden them all over the place without a saddle, holding onto their manes. Webster is the youngest of seven children. He said he loved being active. He climbed trees like a squirrel and hung upside down. His mother would "squeal" and fuss but he kept going. He said he swam like a duck in the river, only coming up for air once or twice. People thought he had drowned but he hadn't. He spoke of his father's general store on the reservation. It was big and carried all sorts of things, including clothes. People from all over came, not just Indians.

Then he looked up at a picture of Jesus he has on his wall. He said, "Jesus is 'the man,' the most powerful. He talks to me just like we are talking. Jesus can give you a new life if you are miserable. There have been times when I was overwhelmed, and Jesus helped me through the hard times." His words were not preachy or hokey. They were loving. I asked how the Indians came to accept Jesus, and he said, "It was long ago darlin'."

I asked when he first noticed girls, first kissed a girl. He smiled that bright smile and said, "It was a long time ago, that's for sure." He told me that I had turned his life around, that he had a new zest for life. Me too, dear Web. Me too.

September 17, 2002 *Turtle*
Autumn coming. River flowing. Smooth green acorns with their bumpy brown caps. Walking all through the woods, ferns, fallen sticks, a few crimson, yelloworange leaves, up the ravines, legs strong. A turtle! in the ravine, bright yellow head poking way out, yellow design on muddy black shell. Happiness to be surprised by the turtle.

September 17, 2002 *"That land of winding waters"*
What is it about this river? I come here and I open. I am mystery. I become. I cry with the purest joy. I write. I am alone with this infinite space. I am washed clean and made whole.

I love that land of winding waters more than all the rest of the world.
    —Chief Joseph, Nez Perce

"She called me Web." Chief Webster was speaking of his wife, Mary, who bore him nine children and cooked many meals and washed many clothes and cleaned house and loved him. His daughter, Shirley Little Dove, told me that she was patient and clean and loving. Carl speaks very lovingly of his mother. He says she was a humble, giving woman. She was always feeding people. I wish I had known her. I met her and Webster a few times many years back but did not get to know them.

So happy to see Mr. Custalow today, and he is happy to see me. He had on a red sweatshirt and grey Nike sweatpants. Carl always makes sure his father has nice clothes. The Chief said he was watching the light on the wall to see it move when I opened the kitchen door to come in. He was in bed with his "pups," awake, shaved, and bathed. We sat close together, and I held his hand, rubbed it. He spoke of talking to Jesus and cried. He said that Jesus was not a "big shot; he talks to everyone. He comes to help me with my dogs. He is all powerful." The Chief said he loved me, loved having me come, that I meant so much to him. It is hard to put into words what this means to me.

We wandered into talking about how he played the guitar and sang. He and a few other men played in Gloucester, Middlesex, West Point, and King and Queen. I had James Taylor on the CD player and he liked it, said he used to have a Gibson and made it have twelve strings instead of six. He lifted his strong, gentle hand and said it had "a charming sound." He said as he recalled, that he invented the twelve strings and that Stella guitar stole his idea and made thousands and "gave me a Stella guitar for my idea." He said, "They have the money, but I have the memory." Carl told me later that he thought it was the "Hoe Down" TV show his father had been on. He said that the guys from the show would come and play with his dad on the porch of the white house on the corner. Everyone would come and listen. They'd play half the night. I asked Carl if there were any tapes of him playing, and he said they didn't even get electricity until around 1954.

Mr. Custalow also told me he invented a lot of things he needed—a way to get the boat in and out of the river, no matter what the tide was doing. He said his wife told him, "Web, you should do something with that [register it and patent it]." But Chief told me "I was too busy to do anything with it," and he smiled. He told me he had a good memory, many memories, and he could "draw himself a picture" of anything and "visualize" it, that he was lucky to have a good memory for his age.

Chief spoke of the sun heating the whole world and "To tell you the truth, man has hurt the earth. He has not taken care of it, and caring for it is why we are here." He was sad about it, and I am too.

My heart is sorry all the time. It cries every day. All the Indian country feels sad.
  —*Big John, Skokomish*

In the kitchen, I brushed his hair; he loves how that feels as long as I don't brush too hard. I hugged him goodbye, and he hugged me and told me he loved me. I said I loved him too. He told me that I was pretty. I never bothered him about a drive. He likes to go, but he has a hard time leaving his pups. I just let my time with him be whatever it is and take whatever it is. I take care of him and show him respect. He takes care of me and shows me respect. I come away with much. A simplicity and part of his spirit touches mine.

September 19, 2002 *"I always loved eagles."*
Chief Webster in his cozy bed, dressed in a black sweatshirt with a wolf on it and black sweatpants—clean and shaved by Carl. He looked so handsome, beautiful to me. I told him he looked nice. We sat in the small room side by side, and he spoke again of Jesus in the same way. He talked about a dream he had of cows, and he brought one home to the green meadow so she could eat. He said that too many visitors put him "out of kilter." Then I said I would put on some James Taylor music and do a little cleaning up. He wanted to "rest his neck" in the bed

and look at his picture of Jesus. I mopped the den and kitchen and did a few dishes and sang with the music. Then I went back to his room and sat again. I asked if he might like to get out today and he said, "Well, to be frank, I'm just enjoying relaxing back here." I said I understood. I added that sometimes you can relax too much, then told him I was going to Carl's for some fish cakes for his lunch. Was he sure he didn't he want to come? No, not now. When I got back, he said to me, "You know, I've been thinking. Relaxing is good, but that's just one thing." I agreed and said there was more than that to life. He agreed heartily. I said that his memories were wonderful (He had just experienced some strong memories about the Wabash Cannonball train wreck, stirred up by a song.) but that he needed to live in the world too. That he could get out and see Jesus' creations. He felt powerful about wanting to get out, that he could not just lie in that room and relax. He seemed renewed with the thought. I finished making fish cakes, cooked apples, and peeled tomatoes, and he came in to eat.

Before we could leave, a visitor came. When I opened the door, the dog, Gomer, escaped, but the visitor rounded him up before the Chief discovered it. We sat in the kitchen, and the two of them reminisced. Chief Little Eagle spoke thoughtfully about his own name to the fellow. "I got my name because I love eagles. Sometimes you have a strong feeling about a certain animal or bird. I always loved eagles. So they called me Little Eagle.

After the visitor left, we drove around and went to the place where Lee and I are building a home on the river. I had not told the Chief we were building next to the reservation. I feel awkward being on "their land." Lee was at the shop he is building, and when the Chief and I left, I told him that we were building there, and was that okay? He said, "Sugar, you do whatever makes you happy."

When we got back, he wanted to know about his truck—where the keys were and how to get it going. He got out, and we made our way to the truck. He opened the door and hunted for his keys. They were in the ashtray—I found them, and he said for me to start it up. I did, and it started right away. He wanted to be sure he knew where the keys were, and he would get his son Leon to clean it up and take him for a drive. We went in and I got him some juice, and he laid down. He wanted to know when I would be back. I told him the weekend and kissed him, and he kissed me.

September 22, 2002 *Pulling out of a dark place*
Stopped by for a visit with the Chief today. When I arrived, Mr. Custalow was anxious. He wanted to

get things straight in his house and make sure his dogs were safe. I heated up some homemade chicken soup, and he came to the kitchen table and sat down. He ate and slowly relaxed some. I held his hand and rubbed his back. I brushed his hair. He wanted music, as if he knew this brought him good focus. It was clear that he did not want me to leave. I felt like he needed me to be there to keep him from falling into that dark place again. How well I know that feeling. After awhile, I told him I would return on Tuesday. He kissed me, and I kissed him and told him I would think about him. He said he would miss me.

September 23, 2002 *At work in town, missing the Chief*
Wondering how the Chief is today. Missing him so. Curious, I called the woman who stays with the Chief on Mondays, Wednesdays, and Fridays. I was checking on him since he was so restless yesterday when I visited. I asked her not to let on as to who I was, and she didn't, but then she said, "He wants to tell you something." I could hear him crying and talking, and he said they were "tears of joy and that he loved me." How did he know it was me?

September 24, 2002 *Impending war, "Attacking a bee's nest"*
Beautiful autumn day, but we stayed in as Mr. Custalow said he was tired and not feeling well. I washed his hair with lots of warm water and brushed it, and he looked and felt better. Then he laid down while I cleaned up, but he didn't sleep. I made him some fried potatoes and chicken soup. I also fixed him sliced bananas with cream and a little sugar. He liked that. We sat in the kitchen and talked. I had mentioned earlier that President Bush wanted to go to war, and he didn't say much at the time except to ask with whom, and I said Iraq in the Middle East. Later after lunch as we sat, he turned to me and asked again about Bush. I told him, and he said that the United States thinks it can beat Iraq, but that all those countries there will join forces and fight against us because they don't like us. He said we have taken too much of the world, right back to how this country started, referring to the Europeans taking nearly everything, including lives, from the Indians who were already here.

It does not require many words to speak the truth. I am tired of talk that comes to nothing. It makes my heart sick when I remember all the good words and all the broken promises. There has been too much talking by men who had no right to talk.
 —Chief Joseph, Nez Perce

The Chief and I

The Chief said those people in the Middle East are not afraid to fight, and for us it will be like attacking a bee's nest. They will all swarm after us. He spoke of September 11, when "two planes flew into buildings in New York City." I told him I agreed with all he said. He said the United States is viewed as a bully by other countries.

When something is scarce...
I then asked him about the eagle he spoke of last Thursday when the visitor was here. He said that one time, he and Carl saw a young eagle flapping its wings, struggling in the river with its talons stuck into a fish that was too big for it. They watched until the eagle got the fish to the shore. He said that they had a place in the woods where they used to raise eagles. Usually, eagles lay two eggs, but sometimes only one would survive. He used to go to West Point and buy a box of assorted fish to feed them. He showed me with his hands how he would cut the scales, clean, and cut the fish for them. I was wondering why he did that for wild creatures, and he explained "when something is scarce, you have to take extra care with it until it comes back strong."

I felt then, and do now as I write, deep emotion. This experience for me is such a gift. There is gentleness, wisdom, and love in Little Eagle. I told him that I was sitting there listening to his words, thinking how he lived his life in love, not hate, not war. He went to lie down, and I got his blankets straight. He said, "When you come back, I may not be feeling real well, but that does not mean I don't love you. Right from when I met you, I loved you." I said I knew and felt the same. That Web express this touches me deeply. I explained how sometimes at work I miss him and look at his picture on the computer— how it was like an electronic book, and you type in something, "Mattaponi Reservation," push a button, and his photo comes up.

Before I left, I asked the Chief how he had known it was me on the phone yesterday. He just smiled.

September 26, 2002 *Talking by the Chief's bedside*
Web wanted to discuss possible war with Iraq. His opinions were the same, but he wanted to discuss it again. He kept saying that war was devastating. People did not, could never understand the consequences. They think they know what will happen, but they can't, and it is always worse than anyone can imagine. There is much suffering. He cried and said, "I have seen so much." Why do people say we should learn from our elders, but we don't? Why do they say we should learn from history, but we don't?

The Chief and I

We had no answers…
We began talking about the goodness of God, and I wondered about abused children and where was God for these children. We had no answers for this. He said that some preachers are really devils. The talk of children triggered memories of the Great Depression and how sad it was for these children, "all skin and bones." He said he gave his last dollar to buy food to give parents. He did not give the money, because sometimes they would buy whiskey and beer with it. The Chief's face was etched in memory and pain, and he cried again as he spoke, the sorrow he felt all those years ago fresh in his heart.

Mr. Edwards' coattails
On another subject, I asked if he went to school on the reservation, and he said yes. He said he liked school "pretty well actually." His teacher came from nearby—Mr. Edwards. In winter when it snowed and the river iced over, Mr. Edwards would ice skate to school on the Mattaponi. At the end of the school day, Mr. Edwards would jump off of Mr. Custalow's wharf onto the frozen river—fifteen feet of frozen ice—and carve his initials in the ice with the toe of his blade, then skate off. Mr. Custalow recalled how the children would watch him with glee as his coattails sailed behind him. They would watch until Mr. Edwards was out of sight.

"A good sign…"
At lunch, I asked if there was any significance in two adult eagles and a juvenile flying in circles and squawking above the site where Lee and I are going to build our house at the river. He paused and said, "It is a good sign, and if they were there, it is a good place." Chief, I love you! I was silly to ask this probably as if he had some magic answer, and he was kind in his response. He was all clean and shaved and looked so handsome.

October 1, 2002 *A dark cloud lifted*
I found out that last Thursday after I left, the Chief made his way to his truck alone, got in it, and honked and blinked lights wanting someone to come help him get it going. Gomer slipped out during this time and came back a few hours later. Poor pup! He just wants to be a dog.

The Chief worries that his dogs don't get enough to eat, and he tries to get them to eat, but sometimes they just aren't hungry. I gently say, I guess they are like you, not hungry right now. The Chief tells me that sometimes he had too much vinegar, and it upset his stomach. So today I said, maybe the dog had too much vinegar. He was still preoccupied though, so I said, "You didn't laugh at my joke." Distractedly, he repeated what I said, and I said, "But you didn't laugh." He looked up at me and focused, then gave me the greatest smile. I told him that sometimes I worry too, so much that I make myself sick, and I have to make myself stop. He said, "Me too!" Then he unblocked the door altogether, saying that if he left the door open, the dogs would come back more easily.

Today, I washed his hair, massaging his head, washed off his back and put aloe lotion on it, brushed his hair, put a clean shirt on him, and played music. He felt much better. I sat next to him and asked about the Harley Davidson he used to have. I held his hand, and he held mine tightly, and he talked. When I had arrived, I told him he had not given me a smile, and he said he didn't feel like smiling, but I smiled at him and he ended up smiling at me so sweetly and sort of funny. Then, when I left, I kissed the top of his head and told him he had not even given me a kiss. He beamed and kissed my cheek. He was relieved to have someone pull him out of himself, out of his troubles. I am grateful when someone does that for me. It is like a dark cloud being lifted. Being with Webster does that for me. I can arrive feeling too much of the world, and then, the two of us just being, makes it all slip away.

October 3, 2002 *Sorrow and joy*
We know that the white man does not understand our ways. One portion of land is the same to him as the next, for he is a stranger who comes in the night and takes from the land whatever he needs. The earth is not his brother, but his enemy—when he has conquered it, he moves on.
　　—Chief Seattle, Suqwamish and Duwamish

How smooth must be the language of the whites, when they can make right look like wrong, and wrong like right.
　　—Black Hawk, Sauk

Day of immense sorrow and heavy heart. For now, the politicians have been successful in getting the reservoir permit granted from General Rhoades of the Army Corps of Engineers in New York. They reversed the comprehensive Norfolk Army Corps' decision in a rubberstamp move. Carl will not speak of it, but I see his face. I cannot bear the thought. Watching the earth die at the hands of greed is a burden that leaves my spirit broken. I think of *Bury My Heart at Wounded Knee* and know the extent and power of evil. I think of what this country was, the incredible beauty of the natural earth and see what it is today. I see us heading to war. And this river that brings solace, that brings me spiritual strength in this world, is now in grave danger. So yes, my heart feels deep despair. How the Indians must have prayed to the Great Spirit, and to what end? As I sit, a heron stands across the river, a messenger from heaven. What is its message?

Nothing the Great Mystery placed in the land of the Indian pleased the white man, and nothing escaped his transforming hand.
    —*Chief Luther Standing Bear, Oglala Sioux*

The greatest object of their lives seems to be to acquire possessions—to be rich. They desire to possess the whole world. For thirty years they tried to entice us to sell our land to them. Finally the soldiers took it by force, and we have been driven away from our beautiful country.
    —*Charles Alexander Eastman's uncle, Santee Sioux*

Today, Little Eagle asked how I was, and I said, "Sad." We hugged. He was so tired. He said when he is tired, he gets "clumsy." Me too. We sat for a bit. He said, "They sell the river for the dollar." He said he spoke with *them* some years ago and told *them* "it was the last clean river here, and they destroy it—God's world— but they do not listen. They talk in 'we.' I am a man and must endure what other men are doing." The weight of that statement is a heavy stone to carry in this life.

I have carried a heavy burden on my back ever since I was a boy. I realized then that we could not hold our own with the white men. We were like deer. They were like grizzly bears. We had a small country. Their country was large. We were contented to let things remain as the Great Spirit made them. They were not, and would change the rivers and mountains if they did not suit them.

—*Chief Joseph, Nez Perce*

I changed his socks and washed his hair and face, then he went to nap. I cleaned up. When he woke up, I went and sat with him. We talked about his swimming again and how he had many "skills" and was like an "acrobat." I wish I could have known the young Webster. I asked how he met his wife. He said he was out riding his motorcycle, an old "Indian" make and drove to the area where Upper Mattaponi lived. Mary was an Upper Mattaponi Indian, and he met her that day. I asked what made him notice her, and he smiled and said, "She was a pretty little thing." He told me that she visits him now, and they talk and he can see her, "just like I see you. She is young." I cried as he spoke. "But then she has to go after awhile." He cried. He said he missed her so much when she died, that "she kept my things straight and did not let anyone bother them." He said his mother came to him some too and a few years back, his father.

He cried again. He said, "My mother had "coal black hair, like yours."

Here we sit, the two of us in a small corner of the world, and there is so much life here, so much love. When I made his lunch, he read the paper. (He does not usually look at the paper when I am there; I think his eyes bother him.) He has a drinking glass with several pairs of large eyeglasses, and he has a couple of favorites. He read about President Bush wanting to go to war and again said it was a mistake, that Bush was just a man, not God, and he could not know the consequences. That Bush did not have all the power, only God. I told him I would rather spend the day with him than with Bush. He smiled again. I said it was like watching the world die, and we had to go along with whatever it brought. I said, "If the world ends, can our spirits be friends?" He said, "Sure they can."

He saw another article in the paper about the *Titanic;* he was born the day it sank. He spoke in great detail about the ship and the accident, how the captain said it could not go down, that the ship was stronger than the iceberg. He said that man thinks he is more powerful than nature, but God showed him that was wrong. He talked about John Jacob Aster who was "the richest man in the world" then, and he had the *Titanic* built and said it could never sink. The Chief said, "You know, I wonder what it was like to have all that money and what he did with it. Most rich people want more and more." He was inside his own train of thought. I watched his hands, his eyes as he spoke. He said, "If I had all that money, I would distribute it and share it." The Chief spoke plainly and from the heart. He was not trying to impress. A wonderful sweetness.

Beautiful day
When he woke from his nap, he had the purest expression on his face and said, "I've been thinking we need to make a sign and paint 'It's a beautiful, sunny day,' on it and set it as you come into the reservation… so people might stop and see it and take notice of the beauty. People are so busy; they go all around, but they don't notice the beauty." His love breaks my heart with joy. I told him we think alike. I want to make his sign with my cobalt paint and sign his name to it— "Little Eagle." He told me to sign it that way.

This is the world today, full of sorrow, the reservoir battle, probable war. But I have my gifts from the Chief. It came to me as he spoke of the spirits that he saw, the spirits I feel here at the river. I know they are here.

October 23, 2002 *An August memory—Swimming in the night river under a sky of white stars*

I think back to the long days of summer, and how one August night I jumped into the river and swam under a sky of white stars. Now it is October with leaves turning again, another season, and always the wonder returns. I have risen out of a dark pit where I had struggled for days, and now I feel it—the light, the joy, the beauty. As free as the wind, the winding river. I belong to no one but the sky, the stars. Alone on the highest riverbank in the wholewideworld. I am a secret in the woods. I am the mystery that never dies.

October 24, 2002 *Only the wind*
I sit in the V of an old leaning tree on a high bank at the top of the world; the wind blows the river, moves in endless motion. A yellow leaf falls onto my jacket. The marsh, the autumn trees across the river wide and far meeting the sky. Only the wind speaks. Only the river.

November 4, 2002 *Happy fiftieth birthday Sweet Pea.*
Fifty years old today. Open to whatever this land and river offer, to whatever they open in me. Setting out for a walk. For my birthday swim?

Walked through the woods, the winding ravines, down the steep path, up the deep gravel road carved out of this ancient forest. Now I sit at the rivertop of the world with the autumn trees, and my eyes follow the bending river and see straight across to forever. The wind gently stirs. I feel it all slipping away, the things that make the heart lose sight. I know again. I am strong again. Can I go back into the world and hold this clarity?

I am on the silent pier, no one in sight. I stand here calling up my nerve, look at the wide brown water. And I jump in. Raw icy cold. Everything in me alive. Mad dash for the ladder. Now wrapped in two towels I sit smiling. One more time? God, the autumn leaves are immense in their beauty.

Yes, I jump in two more times, and my whole body tingles and is rosy. Happy fiftieth birthday Sweet Pea.

November 5, 2002 *Silly, spunky peas in pod*
Day two of being fifty—still good!

Sitting with the Chief. I asked Web if he got married on the reservation, and he said that he and Mary had gone to D.C. for a visit and decided to get married! Earlier, I told the Chief that I turned fifty yesterday. He said I looked younger, not older. Sweet Chief ladies' man! I told him I went swimming in the river, and he thought for a moment, then said, "That was pretty spunky." I told him that eloping was pretty spunky. We smiled at each other, two silly peas in a pod.

November 14, 2002 *"Like moon on water"*
Today is Chief Little Eagle's ninetieth birthday. I baked him a homemade yellow cake with chocolate frosting, brought balloons and a card—one of my special ones—a silkscreen design by Debbie Littledeer Erskine, with stark, black trees, a blue and pink sky. It is a lovely day, sunny and bright, blue skies, sixty-four degrees.

3:30. Walked along the river, fields, woods, ravines. Now I sit at the place we are building our home. Always, *always* I know truth here. The trees across the river in the marsh stand naked now, their trunks bleached pale. Autumn—another season, another cycle. My heart beats; my spirit whispers; my eyes see; my skin is cool. I thank this Earth for Her gifts, Her peace. Yet how to be at peace in a manmade world where hearts and words are often false and bitter. I am at a loss for now, how to earn a living and be true to myself. How I twist and turn in my spirit, my mind. The Dalai Lama said, "I think I am a reflection, like moon on water." That is what I want.

November 15, 2002 *My Jessica's twenty-second birthday*
Postscripts from yesterday—

Looking outside the kitchen window in his back yard, I said to the Chief, "That chestnut is a good climbing tree. I bet you climbed some trees." He smiled his smile and said he climbed every tree he could, "up like a squirrel and down like Tarzan." I know he did. I had a candle on the table, and he said he enjoyed the "flicker." He notices things like I do, and they mean something to him as they do me. His language is clear, from the heart; his words strong.

November 17, 2002 *The muse*
Rainy morning, candle flickering before me. Already walked with the dogs on this autumn day. Leaves have fallen all over the ground, the sky open with trees growing bare. Now showered and ready for the day. I woke early, and in my sleepy mind the words came to my heart, "Follow your muse." Left me with a distinct and strong sense of indeed following my muse. So I shall as if there is no choice. Spoke with my friend yesterday morning, and in her gentle manner, she spoke of visualizing what you want and moving toward it.

November 21, 2002 *Losing the center*
Feeling very tired and foggy. Coffee brewing. I am losing my center, my calm. Need to write. Can't seem to fit it in with work, the Chief, the house, riverhouse,

and kids. There must be some time in which I can focus on my writing, an ongoing project. No one else will honor it as I must— the solitary life of the writer.

Afternoon
Sitting at the riverhouse under construction, spent the morning with the Chief, my sweet friend. So sad and worried I was. After lunch, he began talking about the Chesapeake Corporation and began rambling and repeating himself, not making much sense. He seemed lost. The phone was not working, so I dashed out to use Leon's phone across the street, and the Chief's grandson Todd pulled up. When we got back in, Web had thrown up. My heart was full of worry. We called Carl on Todd's cell phone. He came right away and cleaned him, bathed him, put on fresh clothes. The Chief began to make more sense, and he looked and smelled so nice. I had just washed his hair. He slept while I made chicken soup and organized his cabinets. I kept checking on him. I hope most of all that he will not suffer. I do love that Little Eagle Chief.

November 22, 2002 *"I went home and bloomed."*
One day Web and I were talking about the yearly tribute when the Chief presents an offering of wild game to the governor of Virginia as part of the 1677 treaty, and he said, "I went to the tribute, and they commented on my advanced age. I think they

expected me to go home and fade like a flower, but even though I was old, I was not a flower that faded. I went home and bloomed."

December 3, 2002 *Gingerbread men*
Pleasant sunny morning with the Chief. The pace calms me. Baked gingerbread men cookies, made turkey noodle soup. Web sat at the table while I rolled and cut the cookie shapes next to him. We loved eating them like two children.

Riverhouse
Our home is being built before my eyes, right now. These fine men doing fine work, and I am deeply thankful. I never think this land is mine; we are just borrowing it while we are on this earth.

December 5, 2002 *Winter's messengers*
Sweet Chief asleep with his cap on. God he makes me smile!!

Awake. Went in to sit next to him, hold his hand, and he said, "I am just lying here waiting for the end, wrapping things up. Getting the dogs straight. Sometimes I have a 'smear'—things don't go right. Like everyone. We all have that. Something is troubling us. But if we are calm and trust that God will take care of everything, it all works out." Tears in his eyes. Tears in mine. Was he reading my heart? Speaking from his heart? "Faith," I said. "Yes," he said. Being here is intensely pure. He spoke several times of the end, being ready, waiting for Jesus to help with the ending. Will Little Eagle leave soon? He sleeps a lot. Speaks much of Carl.

Tomorrow, I will take with me this spirit of love. Today, I will write.

At lunch, I mentioned that a bluebird had been perched on the snowy windowsill in the kitchen. He told me that earlier there were several wild turkeys, and they came close to the house. I asked what he made of that, and he said he had some thoughts on that, but that was all he was going to say. Mystery. Messengers?

We were close students of nature. We studied the habits of animals just as you study your books.
—Ohiyesa, Santee Sioux

December 12, 2002 *My name has been Webster.* It feels good to have pen to paper. It felt good to walk a cool, almost winter afternoon across fields, into deep and secret ravines, through old forest, along the riverbank. To pause at the wintry beach and geese huddled across the marsh, to see them midriver and swoosh! wings fluttering, taking flight, the river splashing white.

My walk. I stopped at the rivertop of the world and saw—the Mattaponi Reservation circle, Carl's home, the small white church, the bends of the river, the horizon of sky and treetops. Sound above me. Wings pumping, black feathers, the white head, the yellow beak of the eagle oh so close. A smile from far inside, from a place long ago.

Today, the Chief. I arrived in a bustle of too much on my mind, of Bush's desire for war, of Jimmy Carter's words, "We do not learn about peace by killing each other's children," and of others' deep troubles weighing heavily on me. The Chief asked how I was, and I leaned into him and said I was a mess. He stroked my hair.

Later, we sat side by side in the living room. I was reading his Christmas cards to him, from many friends as close as around the corner and as far as across the ocean. A squirrel was in a nearby tree, and we laughed about the Chief climbing trees like a squirrel and down like Tarzan! I asked about the turkeys he had seen last week. He said they were messengers, and I asked what. He said they were doing their duty that began a year ago. They were telling of the coming of snow. Then he began yawning big, frequent yawns. His words began to slur and become confused. I heard him say, "…letters, words of love, my name has been..." I said, "Little Eagle is your name." I said, "Web. Web is what your wife called you—Web, it's time to eat. Web, your guitar sounds good." I held onto his shoulders, hugged him into me, held his hand gently, all gently, told him he was doing fine. He said, "Yes, that is what my name has been. That is what my wife called me just like you are calling me now." Then his eyes lost focus, his words were lost. I went to get the dishpan and paper towels in case he threw up like last time. He began to heave, and he threw up clear mucous. After that, he began to ease back, his speech returning, his eyes focusing. He told me there was a message inside of him he had been trying to say but he could not, then. I told him some things were between a person and God, and he said, "Yes." I told him too that sometimes our thoughts are not clear and it takes time for them to become so, and when that time is right, the message will come out. He said, "Yes." I helped him to bed for wonderful rest on clean sheets.

December 17, 2002 *Just me*
I walked today and sat up on the high bank where you can see the wholeworld. I felt immensely grateful to be alone. I find myself too connected in my mind to

other people, how I relate to them, how they influence me, how I allow them to influence me. I lose touch with what is inside of myself, just me alone. When I am at the river, it is a place of little thought, just being a part of the grandness of nature.

December 18, 2002 Full moon *A secret place*
I wonder if these words will rest in this notebook, just for me. My secret place like the rivertop of the world. I wonder if the heart is always a secret, hidden place.

December 19, 2002 *Peeling apples with the dark blue sky*
The Chief sleeps soundly. Cutting apples at the kitchen table while outside the warm weather, the sky a deep wide blue. Cutting with this old knife, worn wooden handle, blade sharpened many times over the years. There is comfort in these old kitchen things.

December 20, 2002 *A visitor*
Today, snow on the ground, cold, frigid waters on the river. Reservation buildings behind the Chief's house are crisp and white with red roofs. He is asleep. He is looking so much better, smiling, whole face full of light.

Now, still and quiet. As I look out the kitchen

window, my thought is that I am a visitor here on this reservation, on this earth? Surely yes. What does that mean, to be a visitor? I think it means we care for, we love, we are compassionate, we laugh with the spirit of nature. We speak from the heart.

December 26, 2002 *The dream of the little girl*
Sunshine, cold. Can't even write these words; I am rusty with words, and I know it. "They [kingfisher birds] didn't want to do anything so extraordinary, only want to fly home to their river." The words from a poem by Mary Oliver, and I only want to write my words. An uneasy place inside of me. I just keep writing these words, hoping to find my river.

I am writing at the Chief's. He sleeps much, eats little. Is he leaving soon? Speaks of Jesus coming to visit with him. I come and rub his aching neck, sit by his side. Talk, listen. Am silent. Hear him talk to Jesus and pray as I walk quietly to the washing machine down the hall. "I love you Jesus," he says. I walk away, leaving him to his privacy. I move about the house, sing, cook apples. The house smells of cooked apples.

I came early Christmas morning, before the frantic pace of the day began, to say hello, get a hug, a smile. *The dream of the little girl*—The week of Christmas, the Chief told me of a dream he had about a little girl who needed a Christmas dress, shoes, a hat. She had golden hair, and he was worried that she would not have something pretty to wear. He was not sure who she was. Today I asked if he had dreamed of her again, and he said no, but that he was very worried about her. He wants to take care of children, just like in the Depression. I am sure that is what this dream meant, his love for children. I told him I thought she probably got a dress, shoes, hat, and a baby doll. He had wanted her to have that too as a gift. Later, he saw a girl at church, maybe the girl in his dream I said. Yes, he said, wanting to believe, both of us wanting to believe.

2003 ~ This Wild Place

January 9, 2003 *Dreams and visions*
Just me in this world, sitting outside the Chief's house while he sleeps a little. Warm and sunny like early spring or autumn.

When I arrived today, the Chief was in good spirits—he had his spunky smile when I sat next to him with cold hands that he warmed for me in his hands. Yet he was troubled with dreams and visions. Dreamt of his mother, father, and Jesus—always Jesus he says, and he felt unsettled with all of it. He said he had a lot of stress trying to sort things through. He said he had many dreams and visions. Carl told me that his father has many visions. The Chief and I talked about these things and how life changes. He spoke of the many things he has lost—family, physical ability. He said, "You know who came again, the one who never leaves me—Jesus."

January 18, 2003 *To be needed*
Frigid temperatures and snow—full moon tonight. The sun is coming through the trees across the pond.

The things you count on in life that bring internal order.

Soon my Tuesdays and Thursdays with the Chief will end. His daughter, Delores, will be staying with him Monday through Friday, as Carl had initially wanted one full-time person. Now with three people coming and assorted children, the Chief gets "out of kilter."

I feel such sadness at the thought of not being with him, not caring for him. Carl asked if I would be willing to come on weekends/weeknights if the need arises. Of course, I will. Need—such a human desire to be needed. I convey to the Chief my need for him. And I know he feels the same for me. I can see it in the light of his face, his gentle smile. Love radiates from him. I must hold to the thought that I will be with him and not let the sadness hurt me. This is surely what is best for Webster, to have his daughter care for him.

Thursday was a good, good day with the Chief. When I walked in and tapped on his door, his face was as sweet as a child's when he saw me. I sat by his side, and he was hungry he told me. Tired of oatmeal and soft things (He has trouble with his teeth, so he has to eat softer foods.). He wanted something like a fish sandwich! I called to see if Carl was home and got a big piece of rockfish. Flour/cornmeal/Old Bay and cooked it in olive oil. Fixed a potato, mashed it up with butter. Pineapple, bread, and cocoa. He loved it! Came right to the kitchen. After he ate, I washed his face and put aloe cream on it and washed his hair and brushed it. "That brush is a good thing. It feels good on the scalp, as long as you don't brush too hard." I think one time I brushed too hard. I guess that was his gentle way of making sure I did not hurt him.

He wanted to listen to songs that brought back memories—Ocracoke Songs. We talked about all the bluebirds, and he said, "You know why they're here, don't you?" "Snow is coming," I said. He told me about a bird that had hit the window a few days before, "Bam a lam!" "Bam a lam?" I asked, and he said yes and smiled. My dear sweet friend.

January 28, 2003 *"I do not think the American children want to hurt me."*
Tuesday. Spent the morning with Chief Webster Little Eagle Custalow. I have one more day, Thursday. Monday evening, I was very anxious. I think it was knowing my time with him would soon end. Seems silly in a way, but not really. Something as deep as that river has happened between us. A bond of caring for one another. A mutual sense of what it is to live in this world, to be a human being.

We share language like a dance. We are friends. Of course, I will see him, but it will be different. They will be visits.

Today I told him about a little Iraqi girl I saw on the news, a beautiful child with a bright smile and shining black hair. She said, "I do not think the American children want to hurt me." And she smiled as if this thought was true and she would be safe. All I could see was her smile gone and fear and pain and dirt, not her pretty dress and smile in the green field where she spoke. I told Webster that I thought of him when I saw the girl because of his love for children. He said, "Yes, we must love children. Jesus said we must care for them, take time with them. They are young, and they they look to adults to help them." He had tears and so did I.

January 30, 2003 Last Chief Day *How do we learn about peace?*
A surprise dusting of snow. Candle lit. Have slept poorly for several nights, and I feel edginess creeping into my brain. Coffee. Solitude. Poetry. Writing. I feel a small soothing.

My mind is a blur and a whirl of thoughts and recalled conversations. The little girl in Iraq, Bush's war, Jimmy Carter's words about peace. How do we learn about peace? About love? We learn by living them the best we can…

How grateful I am for the river, the stars, the mystery. Oh Sweet Pea, you are so tired, such a tearful mess, but you will be fine. Wear your strong heart. Web and I did not make a big deal out of this last day. We did the same things. We hugged, and I told him I would be back.

February 3, 2003 *My sweet friend*
Yesterday, I visited the Chief, my sweet friend. I sat next to him as I often do and held his hand as I do. I brought up the space shuttle Columbia explosion, and he knew about it, had watched on television. Right away, with conviction, he said, "They had no business going up there." Do you see why I love this man?

February 8, 2003 *Missing each other*
I miss being with Chief Little Eagle. I went yesterday. I had told his daughter Delores that I would try to come Thursday, but I ran out of time. A haircut appointment lasted ninety minutes because the hairdresser talked so much! I was so frustrated, and out of character, I did not call the Chief to say I would not be coming. I don't know what I was thinking. Then yesterday, I dropped in without calling

first. The Chief was having lunch, and when I entered the kitchen, his face crumpled into tears, then mine. I had thought that maybe he would not miss me but so much as long as he had his daughter with him. I felt awful that I had not called yesterday to say I would not be there, but my heart was so happy to know how much he had missed me. I have never let him down. We hugged and kissed cheeks, and I sat with him and held his hand. We talked.

Delores showed me all the things she had done, cleaning, arranging, new curtains, new foam for the sofa cushions. She is a warm, loving woman. She told me how she used to climb the highest tree on the reservation and see the river. She would watch for her father as he came in from fishing, then run to meet him to see what he had caught. She said she fished with him into the night, then would go to school the next day on the reservation. (Later, like the other First American children, she had to go away to school, and she said she was so homesick. Back then, if Indians wanted to get a high school education, they had to go to other states like North Carolina or Oklahoma. In Virginia, they were not recognized as Indians. It seems the government simply wanted them not to be, at all.)* Her father taught her to drive the tractor, and she would hand him tools when he would be under his truck or tractor fixing it. Through Delores' eyes,

I saw a side of the Chief I had not seen before, as a father to his young children.

Sitting with the Chief again, he turned to me and asked me what I thought of this war. I think he misses our conversations, just like I do. I told him we had not tried to explore peaceful resolutions. He said he thought Bush had gone to war for Iraq's oil. He said he thought the world was near its end. He said that man was destroying this earth. I rubbed his back and brushed his hair and promised to come back Sunday. I may come back again today while the realtor shows our house. I can never harm the bond we have. I am so blessed with his love. I also mentioned that we were still fighting the reservoir, and he said we could at least try to save part of the earth. Exactly my thoughts.

Walter Ashby Plecker was the first registrar of Virginia's Bureau of Vital Statistics, which records births, marriages and deaths. He accepted the job in 1912. For the next thirty-four years, he led the effort to purify the white race in Virginia by forcing Indians and other nonwhites to classify themselves as blacks. It amounted to bureaucratic genocide.—Warren Fiske, The Virginian Pilot, *August 18, 2004*

February 22, 2003 *The Chief's Special Day*
With much snow and changed schedules for the Chief

and me, it had been at least a week since I had seen him. I set aside this morning to go over to his house, fix him breakfast, and visit—just the two of us again.

When I arrived at 9:00, everything was clean and quiet. I walked to his room where his door was slightly ajar and tapped. The Chief's eyes were open, and when he saw me, the first thing he said was, "Quick, come in and turn around." He pointed to the wall facing his bed, the one with his beloved picture of Jesus.

I could see in his face that something significant was taking place—his dark eyes were shining and focused forward. I went in and kneeled beside him. He said that Jesus had just been with him and had delivered a beautiful sermon and had sung hymns to him. He was so happy I had arrived and hoped I would see what he had seen.

The Chief told me that Jesus had come all the way from heaven, from His Father's Throne, not simply from around the corner on the reservation. He often said that Jesus was not a "big shot," that even though He was the most powerful man in the world, He was loving and kind to everyone.

The Chief said that he had been wanting to get to church but could not, and Jesus came to deliver a

special sermon for him. He knew it had been an
experience that few people had, and he was certain
that he would experience it only once in all his years.

His face radiated purely with great joy and love,
and he wept. He had dearly wanted someone to see
what he had seen. I said I thought Jesus had not
wanted anyone else there, that he had chosen Little
Eagle alone to receive this Holy gift. I told him that
I had not witnessed what he had, but I was certain
of the joy I saw in him, and I believed that what he
described had taken place. Still, the Chief wished
someone could have seen, that he could have a
photograph maybe. I said that what happened was
in his heart, a place far more sacred and true than a
photograph. He was good with that.

He also told me that before Jesus came, his wife Mary
had been next to him in the bed. She was cold, so he
covered her with a blanket. But when Jesus came, he
was alone.

For nearly an hour, we sat together, and he spoke of
Jesus' sermon and His hymns, more beautiful than
any he had ever heard. He struggled to say how much
he loved Jesus. I told him I thought some things were
too powerful for words.

The Chief was intent on not disrupting the mood or losing the significance of what had happened. He spoke passionately, his dark eyes full of light, saying that Jesus was only about love, and he loved Him more than anything. Jesus had brought him into this world and had loved and taken care of him his whole life. He used his beautiful hands as he spoke, still hoping to find words to say how deeply he loved Jesus. I asked if maybe he loved Him more than all the stars. He smiled at me with that Web smile and said yes.

He seemed at peace to have shared his experience with me, and he grew tired and wanted to sleep. He was too full of emotion to eat, so I let him be. I kissed him, and he kissed me. I checked his feet to make sure they were straight and even, and we both laughed. The doctor had told him a few months earlier that he needed to keep his legs level, and checking his feet had become our playful ritual.

I told him I would record this event in my journal. He wanted that very much. Next time I visit, I will read it to him.

March 2, 2003 *Looking back on his life*
Stayed with Little Eagle this morning. What a smile when I walked into his kitchen. Carl was bathing him

(top part!). The three of us visited, but the Chief was not hungry. He said he was feeling "all flunky."
He took a nap, and I went to our river house to take a load of packed boxes to store in the space above Lee's workshop. When I went back to the Chief's, I read him my journal, February 22, 2003, the day Jesus came to him. He listened intently and was so pleased. It made me feel good. I told him I would write it up and give him a copy. He wanted that. He recounted the experience again and told me how looking back on his life, he now realized that all the hard work he had done, he had done through Jesus. "I am not a big man, and I did a lot of hard work, so Jesus must have been there to help me." As I prepared to leave, I told him I had to go pack more boxes. He said I had to go do my work, that he respected that. He spoke of the special bond we have. It was important for him to tell me what my visit meant. He said, "You came here today to extend your love. Not many people take the time to do that." Chief, how could I not? I am always amazed at how he expresses himself, the language he uses. He speaks from his heart. And he never makes you feel guilty for leaving. He is appreciative and gracious.

March 11, 2003 *All around me*
I see far across the true sky, the deep river oh so brown and watery, my tree, my tree holds me high enough to

see it all, to feel the spirits reaching for me. These are in my heart. Tears of joy, tears of yearning. Tired tears. Early to bed but oh so weary. Not feeling well. I miss my crazy words, the world of green, of soil, of the smells so deep I ache with joy. So deep I know there are spirits all around me.

March 21, 2003 *First day of spring*
Chief Little Eagle left this earthly world this morning around 2:30 a.m. Todd phoned, and immediately I asked why he had called, fear gathering in my stomach. He said, "Pop Pop died this morning." My heart…my heart broke. With love, with sorrow. I could not stop the tears, the hurt. How I love that man. No words to express it, just like he said of his love for Jesus. I quietly asked Todd if I could come over, and he gently said maybe later in the afternoon, that there was so much family now. I spent the next hour not knowing what to do with myself. Then I could not stand it. I called Carl and asked if he would please do me a favor, please give your father a kiss, tell him I love him, and 'straighten his feet' for me. I made him promise. He promised, then he said, "Why don't you come yourself. You were so close." What relief I felt.

I arrived and went into the Chief's house. The children were all at the kitchen table discussing the business of death. I went into his room—my last visit—and he looked like he was asleep. I kneeled next to him, kissed his head, held his hand, rubbed his arm and spoke to him—I love you, now you are with Jesus, your shoulder no longer hurts, you can breathe without trouble, you are with Mary, your parents. You are swimming, climbing trees, shad fishing. I rested my head on his shoulder. I touched his beloved picture of Jesus, then touched his face. Before I left, I straightened his feet, and I placed a copy of my journal entry about his visit from Jesus next to him. I was happy I had had a chance to read it to him a few days earlier. I touched my nose to his and kissed it like I always did.

Peaceful heart. I thanked Carl. He said his dad would have wanted me there. He said, "His eyes were closed, but he saw you."

March 22, 2003 *Moving*
Tomorrow we are moving from our St. Stephens Church house. We are staying at a friend's garage apartment until our house is ready. I know it must be hard for Lee to leave the house he built by himself. Now we are building a home of our own on the river. This time, Lee is doing the stonework outside, cabinets, and tiling. He is creating a handcrafted fireplace like he did in his other house, but this time

with layers of shark's teeth and arrowheads from the river banks, stones, shells, and fish imprints.

March 23, 2003 *Moonlight and Little Eagle*
First night in the garage apartment, surrounded by fields and woods.

And this... In the moonlit night, I woke from a deep sleep. I woke alert and clear minded. I woke in a peaceful state. I felt the loving spirit of Little Eagle enter my own spirit, strong and true. Right into my heart. I physically felt it. How can I describe it? It was an energy, not a touch, a beautiful energy of love. I felt great joy, great calm, certainty. He has become a part of me. No words to hold this.

March 26, 2003 *Regalia ~ leather, feathers, and beads*
Breathe first. A wild ride these past days, packing, moving, the Chief's passing to another place. I smile because I see him free of pain.

I have had little sleep. Today, there is some respite. Going to the St. Stephens house to finish packing and cleaning. And I am writing. It came to me in the pre-dawn that I want to get back to language in my journal, not just words on the page. Of course, that is all I have had time for. Many new revelations.

The Chief and I

Spring—how perfect.
I begin with the Chief. He came to me in the night. His presence, his pure love became part of me—into my heart, into my spirit. He told me early on that I had turned his life around—and he mine. I believe our spirits are woven together. More than anyone who has died, I know he is not gone. He was a special gift of love God gave this world.

Yesterday at work, I sat looking at his photo, his ninetieth birthday with the cake I baked him, and again something strong and clear came over me. I knew I had to quit the evening teaching job and write this journal. Last night, I dreamed Webster's hand reached for me. He placed it on my arm. His hand and arm were strong and young. I felt great comfort, and doubts were dispelled.

At the Chief's viewing Sunday night, he looked beautiful in his regalia of leather, feathers, and beads. I touched his sweet nose again. I wanted, had to go, even though we had moved all day and I was weary to my bones. I wanted to be there for Carl, too. He wears his eagle's gaze, but I know his heart is filled with sorrow.

Delores enfolded me in her arms and stood with me by her father. She told me the night he died around 2:00 a.m., she was with him and he was radiating with joy. He asked her if she saw. Jesus had just been with him and told him his mission on earth was now over, that Delores and his dogs would be fine. The Chief told Delores there was a message on the wall, asked if she could read it. He tried his flashlight, but it was dead (I found out later that Carl had just put in new batteries a few days earlier.). She told him the message was that Jesus loved the Chief very much and he would be fine. He died at 2:30. I believe Jesus was there with him.

I talked with Carl that evening on the phone about many things. And I knew, again, that no matter what path we take, what actions we choose, whatever our personality, we must strive to be loving and kind above all else. These must direct us, otherwise we spin out of balance. We become fearful, lost, or angry. This was indeed the Chief's legacy. This was his life, the gift he gave in this world.

The day before the Chief's funeral, I had a flash of an idea about the sign he had wanted to make a while back—"'It's a beautiful, sunny day,' and set it as you come into the reservation… so people might stop and see it and take notice of the beauty. People are so busy; they go all around, but they don't notice the beauty." I wanted to make the sign before his funeral. I ran it by Carl first, then I spoke with Curtis Mason, who makes signs, and explained. He had one made by the next morning and would not take payment for it. Lee put the sign up as you enter the reservation. The Chief has his sign.

Little Eagle's family asked me to read my journal entry, "The Chief's Special Day" at his funeral. They found it next to the Chief after I left the day he died. As I sat there at the funeral service listening to others speak and sing, I started getting anxious. I decided as I sat in the pew, that this would be a visit to the Chief, like walking into his home, calling out his name, sitting with him, holding his hand, and reading to him. This calmed me. My voice faltered only once when I got to the part where "I checked his feet one last time." I added this: "I had the opportunity to read this to Web, and Friday when he had gone to be with Jesus, I lay a copy by his side and 'checked his feet' one last time. I have to say that I often thought how much like Jesus the Chief was. He was so kind and loving, and he was surely not a 'big shot.'"

March 29, 2003 *Chief Chief*
Carl is catching shad like crazy, two-hundred then three-hundred in a short period of time. I told him his father must be guiding him. He laughed and said, "Hell, he's in the boat with me!" Chief Chief.

April 4, 2003 *A reprieve*
I do feel a lot of loss right now. I could always turn to the Chief for a reprieve from the world, a place to go and feel his hand touch my hair, or I would rub his hand. We laughed about silly things that no one else cared about. I keep telling myself he is in my heart, but I miss him.

April 13, 2003 *Coming to me*
The Chief's spirit is with me at river. The sense came to me clearly.

April 25, 2003 *Mattaponi River, house site ~ transitions*
Raining. Sitting in my car, a canopy of green trees encloses me. Hard to be here, yet not to visit the Chief. We were two silly peas in a pod, away from the rest of the world. I walked his dogs today. Since I have not been with Web, I do not write in my journal much. Being with him created a still place that led me back to myself. I know he came to me two nights after he left. It is as certain and clear now as it was then. But I miss his company, his smile.

So much has happened this past month. Web's passing, moving into the garage apartment, being sick, teaching the alternative education students at the high school, then deciding to stop, and the all consuming fight to save the river. I dream about the reservoir battle at night, wake up tired and worried, stomach filled with stones. Another three and a half weeks until the next Virginia Marine Resource Commission (VMRC) hearing. Preparing for the move into the new house, my daughter graduating from college, my son graduating from high school. Transitions. I will emerge into new places, both internally and externally.

May 2, 2003 *"The soft sound of the wind darting over the face of the pond"*
I am sitting outside of a bagel shop in a strip mall. Cement, development after development, phone wires, cables. I feel anxious here. Want to go back to the river. I am not meant for this place.

The sight of your cities pains the eyes of the red man. But perhaps it is because the red man is savage and does not understand. There is no quiet place in the white man's cities, no place to hear the leaves of spring or the rustle of the insects' wings. Perhaps it is because I am a savage and do not understand, but the clatter only seems to insult my ears. The Indian prefers the soft sound of the wind darting over the face of the pond, the smell of the wind itself cleansed by a midday rain, or scented with pinon pine. The air is precious to

the red man, for all things share the same breath—the animals, the trees, the man. Like a man who has been dying for many days, a man in your city is numb to the stench.
   — *Chief Seattle, Suqwamish and Duwamish*

May 6, 2003 *What I have of love*
The wide page stretches before me like the sky above the river, reaching to touch the small mysteries of my heart. What I have of love of love of love brings me to my knees trembling to kiss this earth, the soft moss beneath my cheek.

It was good for the skin to touch the earth, and the old people liked to remove their moccasins and walk with bare feet on the sacred earth. ... The soil was soothing, strengthening, cleansing, and healing. ... For the Indian to sit or lie upon the ground is to be able to think more deeply and to feel more keenly; he can see more clearly into the mysteries of life and come closer in kinship to other lives about him.
   — *Chief Luther Standing Bear, Teton Sioux*

May 8, 2003 *A vision*
Garage apartment
I went inside to lie down a few minutes ago, and in my half sleep, I saw a clear vision of long-ago Indians,

bare-chested, paint on their faces, springing from the old forest, the ferns, the green. They were men. I don't know if women and children followed; I wondered about them. They were at the place we are building our home, in the forest I call the passage of spirits, a place I have especially felt from the very first time I went there, that there were souls from the past. It was clear to me that this was a vision, not a dream; it was the quality of the experience, the clarity, the message. The Indians made me happy, my spirit bright as if I were greeting them. I pray with my entire being that I may bear them into my heart, that they may love who I am.

And when the last red man shall have perished, and the memory of my tribe shall have become a myth among white men, these shores will swarm with the invisible dead of my tribe. And when our children's children think themselves alone in the field, the store, the shop, upon the highway, or in the silence of the pathless woods, they will not be alone. ... They will throng with the returning hosts that once filled and still love this beautiful land. The white man will never be alone. Let him deal just and kindly with my people, for the dead are not powerless. Dead, did I say? There is no death, only a change of worlds.
　　— *Chief Seattle, Suqwamish and Duwamish*

May 15, 2003 *"I do not own one leaf on one tree."*
The reservoir permit was denied yesterday. The river runs wild and free. I saw you Web in my mind's eye, standing at the podium, strong and kind, speaking from your heart, your hands gesturing, your dark eyes shining.

I spoke at the hearing, and at the end I added, "As my beloved friend, Chief Webster Custalow said to me, 'Man thinks he owns the earth, but I tell them I do not own one leaf on one tree.' I ask that you remember his words." I had not planned on speaking of the Chief, only because I wondered if it was my place to do so. I wanted to with all my heart and knew his words should be heard. He began this phase of the battle in 1996. At the last moment, his words came to me, and I had no doubt that I would speak them. Funny, at his funeral, I only faltered at the end. When I spoke of him at the hearing, I was filled with emotion and close to tears.

May 17, 2003 *Warriors*
My first visit to the river yesterday since the reservoir permit was denied. Joy to behold the whole of her. Windy day and small whitecaps, greenness on the banks, in the marshes, the sky full of blues and white. On the bank, two eagles flew above me, and one dove straight into the river and came up with a fish in its talons.

And this... I took a walk, and as I made my way along the length of the riverbank, it came to me! The vision I had on May 8—the Indians rising from the earth. They were rising from the earth to fight for the river. The reservoir hearing was to follow in exactly one week, May 15. I remember thinking the day I had the vision, that the men were warriors. (Maybe that is why there were no women and children.) They were dignified. They were strong. I felt awe, and they were beautiful to me. Yesterday, I knew why they had come.

Now I am thinking of Webster Little Eagle, and I see him young and handsome and strong, dark eyes full of light. Always, I hear the Chief's words, his spirit echoing what is in my heart.

July 20, 2003 ~*Chief Lone Eagle*~
Why should you take by force from us that which you can obtain by love?
—*King Wahunsonacook, Powhatan*

Carl Lone Eagle Custalow was voted Chief Friday night, July 18. He said life has been good to him; he has attained a good education, and he wants to give back. Amazing when you think of what has been taken, when you think of Newport News' fifteen-year attempt to take this river. Just like his father in that

way. He has a vision for things he wants to do, and
I hear a strength in his voice. I see Carl as an Indian
who is his strong heritage, yet who also has been hurt
by the white world. He walks in both worlds and lives
true to himself. I watch and learn as he fights this
reservoir battle with dignity.

A Man of the River

I began long before my birth.

I was already a man of the river,
a fisherman in the dark morning,
pulling in nets of shad by firelight.

I was already a man of the soil,
growing corn and greens
that made my eyes bright, my skin smooth.

I hunted deer, turkeys, and geese
and praised the Great Spirit for their lives.
I ate them, and they made my bones strong, my heart brave.

I watched my father and learned.
We worked side by side,
and I honored him.

The Chief and I

My mother was the humble Earth,
the hands that fed those
who sat at her table.

I began long before my birth.
I was already a warrior.
I stood straight and silent,
but my eyes watched.
I was a father.
My sons watched me and learned.
We worked side by side.

I am another season of those who came before me.
Their spirits whisper in the wind,
their voices sing along the waters.

When I think of my life, I cannot separate it from the
river, the soil, the vast sky
that gave my father, my mother, their parents
       and theirs and theirs and theirs
a sense of what it is to be a part of this Earth.

Today I see that people are separate from the Earth.
The Earth is theirs to be overpowered and used up.

But I am a man of the river.
I am Lone Eagle.

July 25, 2003 *Wings pumping*
Late afternoon Mattaponi River
End of the day. I sit with a small breeze on my skin, birds in the forest, river before me. Green marsh, blue skies, white clouds. I can hear my spirit without the clutter of the busy world. Wings pumping. I look up. An eagle across the open sky. Is it maybe a descendent of one Little Eagle raised? His beautiful hands. I can see them strong, feeding the young eagles.

August 16, 2003 *Justice in America?*
It seems as if many issues have come to my attention recently. It began when a neighbor described the poverty he saw on the Navajo reservation. I had heard things but did not really know details. The incredible nature of it all made it hard to grasp, so I looked it up and found this sampling:

Testimony of Mark Maryboy, Chairperson Navajo Nation Council, Transportation and Community Development Committee, addressing Congress—
Chairman Ney..., as the 2000 Census revealed, ... families on the Navajo Nation commonly live in homes with fifteen to twenty people under one roof, in a house without drinking water, without a toilet, without showers and sometimes even without a kitchen. The sole access routes leading to these homes are generally dirt roads that are frequently made impassable by mud and sand dunes. These are the same roads that over half of all

The Chief and I

Navajo families must travel to haul drinking water twice a week just to meet their most basic health and sanitation needs.

I could not stop thinking about this prosperous country and the lies that perpetuate such inexcusable and obscene conditions for First Americans. Far too many people in our society are so removed from what true human beings can be and are meant to be. I feel enormous sorrow and loss. I know people claim this is the past, but it is not. Until the United States acknowledges and rectifies its past and its present, it is very much the fact of today. Just as with Newport News' schemes to take this river, it is lies and greed and taking whatever they want at whatever cost.

There is too much war and violence in the world, endless use of resources, and environmental rape. An article I read in *Vanity Fair* ("George W. Bush vs. the Environment," by Michael Shnayerson; September, 2003.) was very scary about the systematic and planned methods that are established in the Bush administration to destroy the environment. This is not a democracy; it is truly no different than what happened two hundred to four hundred years ago. The Department of Interior, under Bush's administration, is orchestrating scandals against the Blackfeet, and now the Navajo, of stealing, lying, and covering up. The damage of the past and the same behaviors are very much alive, and I simply cannot—not—see it, feel it. It is wildly out of control, and it is all around. I wonder if most people simply do not see it, are numb to it, or just don't care. I tell myself to be a light, to stand up for what it right, but it is like swimming against a very strong current. I hear the Chief's words echoing what is in my heart. It's crazy, my kindred spirit is indeed a spirit.

I called the Judge, Royce Lamberth, in the Eloise Cobell case. Ms. Cobell is the Blackfeet Indian who has been in court for seven years trying to get the billions owed to her people. The judge has stood up to the Department of Interior's Gale Norton, et al and has won the case thus far for them. I spoke with Judge Royce's legal assistant, Katrina, a young woman with a gentle, kind manner and briefly explained our case. I expressed my respect and appreciation for the judge and wondered if he could offer any advice for the reservoir battle. Katrina relayed my message to him and called back that same day to say the judge appreciated my kind comments very much "as they do not come to him all that often and he is very grateful for them!" She laughed. It surprises me that he does not get praise, but then again why should it? He said he was sorry, but unfortunately he could not comment on the case as it would be improper to do so, but he wished me luck. He also said that our case could end up in his court and for that reason he could not be involved. I sent him a thank you letter yesterday. People like Eloise Cobell and Judge Lamberth give me hope. They are my heroes.

September 18, 2003 *Hurricane Isabel*
I am staying at the river house alone. Well, am I alone? I have the Chief's beautiful photo with me, a candle flickering before him. And who knows if the spirits of my vision are with me?

Big BIG blustery gustery winds. Woosh, trees swaying and bending all night. Bam a lam! Trees falling. And what else? I went swimming in late afternoon before the hurricane arrived. It was very windy, and there were small white caps.

September 20, 2003 *Another world*
When I woke up Friday morning (after a very uncomfortable night's sleep on the floor, sore neck and back; we have not moved in yet), I was getting ready to walk around and see the damage from Isabel, when Brenda from next door (my only immediate neighbor) came over. It was around 8:30. She went back to her house and made coffee at my urgent request! The morning was calm and crisp with blue skies. We made our way down our driveway and could see trees down. We had so many huge trees fallen;

it looked like another world. A forest of fallen trees with their huge root systems overturned. We had to switch to her driveway to walk (She had only had one section blocked, and it was by our tall tall tree.). I was feeling awe I think, awe and fascination at nature's power. Destruction and beauty... It was later at day's end that I was overwhelmed and weary and cried. We made our way back and began the aftermath of the storm—cleaning up. Lee was clearing his way out of the garage apartment where we are staying. They all arrived around 1:00; Michael and Warren came too. In two hours, they had the driveway passable. I thought it would take a week. On the way back to the garage apartment, as I took in the destruction from the hurricane, I cried. It occurred to me that I had not heard crashing in the night. Carl called tonight, and when I told him, he said the trees fell gradually. He said in all his years, he had not seen such a storm. As I walked on Friday by the riverbank, I found green acorns with their bumpy caps and it made me smile to remember last year, the same.

September 26, 2003 *Finding our way back*
The hurricane was one week ago. Trees down all over. I have been edgy off and on. A natural disaster; I know this, and we pick up the pieces. Others in the world have much worse. Still I accept my feelings and try to be positive, try to re-order. We will not move

this weekend as planned. Maybe in two weeks we will have electricity over there (still none here at the garage apartment). For now, I am finishing what I can at the house without power. I am walking. I made a new dog friend, sort of. He is shy but seems sweet and bright. Seems almost like a gentle wolf. I look forward to seeing him.

The saddest part for me is that the places in the forest where I felt the mystery, the sacredness, the spirits… are torn apart. I have been walking amidst all of the fallen trees seeking what I felt before, feeling the loss. Yesterday, briefly, I felt the good again. Maybe it will take some time for the spirits to find their places again, just like me. Maybe we can do it together.

I feel too removed from myself, too many outward obligations. I need to find my way back.

I swam today, chilly but wonderful. The lavender sky reflected in the river. A heron took flight, big and loud and perched in a tree across the river. I miss my times in the big rivertop-of-the- world tree. Someone bought that land and cut all of the beautiful old trees, and it makes me sad to go up there.

September 27, 2003 *Dream of Web*
Sleepless night. Tense and achy. 6:00 a.m. took two

herbal calming pills and slept until 7:30. Dreamt of Web. I was preparing him good food and getting him nice, clean clothes. He rested himself next to me. Thinking of it now, he was helping to calm me as he did when he was here.

September 28, 2003 *I wonder*
We got permission from the county to move our furniture yesterday, and the electricity came on to boot. Carl, Todd, and Chad insisted on helping us with their truck and trailer until the job was done. Todd said we were the luckiest people he knew to have so many trees come down and not one on the house. I wonder if the Chief, Carl's earlier blessing, and the spirits protected me.

October 4, 2003 River *A place to begin*
The morning sun is beaming into the breakfast area, through opened glass French doors. New house, sitting at the breakfast table, Nana's hand-stitched tablecloth—muslin with green leaves and lavender grapes. Blackbirds swarming to the tall trees, eating red berries. My first night here (alone) since we placed our furniture, got water (cold), heat (upstairs). The beauty right now calms me. That is all that matters. Moving stirs up so much physically and emotionally. Throws a person "out of kilter." I hardly know where

to begin. I suppose "doing" is the most important, but I begin with my writing, the river, the trees, thoughts of Web, the spirits. This is the best place to begin.

Hoping to write every day. It must be what I need because I am now crying as if I have found an answer to all of this tumult... as if I did not already come to this answer many times before. I brought my beautiful photo of Web with me last night. And my journal.

October 9, 2003 *How many before me?*
Full harvest moon tomorrow. First night in the river house with certificate of occupancy. Feels like it will be my home now; I can stay. The Chief is with me—his photo, two feathers, one tipped in gold, I found on a beautiful walk, the moonlight all around. I wonder when I look at the moon, how many before me watched this moon, light falling into deep ravines, falling across the river.

I watched a heron walking in shallow water today. Funny how they step and goose their necks along. I visited the Chief's grave today, touched his photo on the headstone, and felt love and relief well inside of me, recalled the night he came to me.

October 10, 2003 *Full Harvest Moon - finding treasures*
Walked and found treasures. A feather (another tipped in gold), an acorn, a shell, a branch with orange persimmons on it, and a perfect blue morning glory now in a white cup on the table.

October 12, 2003 *My new home*
What a beautiful day. Sunshine, warm, blue blue skies and blue water and now the moon hanging over the river. We got lots done. Out of the garage apartment!! I cleaned and am so tired. We have made great headway here, but there is still much to do. Friday night, I got locked out of the house while Lee was with the builder and ended up having dinner with Carl and his family. I had a little wine on an empty stomach and tired body, and I got very tipsy. It was fun except for the headache later.

Saturday, friends came over after my long day. We fixed lasagna. So good to have them over. Anne brought a beautiful framed watercolour she painted of her front field with three dogs. Love it. Michael and his girlfriend stayed for two days. I baked banana bread tonight. Home cooking. I LOVE it here. Walked and cried. The spirits, I felt them and the Chief. 11:30 p.m. off to bed, to my job in the morning—from my new home for the first time.

October 18, 2003 *Amazement*
Happy Birthday Dad! Eighty-two years.
Animals and birds I've seen in the last few days: a fox, a deer, and an eagle right above my head. Heard its great pumping and looked into the trees to see it flying off. The autumn leaves are slowly turning colours. The soybean fields are wide and yellow yielding into woods and sky. The river is glistening and flows forever it seems. Being here is magic. The half moon is above me as I sit at the kitchen table. Tall tall trees tipping the sky. I feel amazed much of the time. A few nights ago, when the full harvest moon was waning just a little, I saw its reflection on the river and walked to the riverbank to see. The moon was misty. The reflection was misty. It was a soft yellow. And the stars were reflected on the river. A heron called out. A BIG splash, then several more spaced out in minutes. Ahhhh! A beaver. I smile. Watch the river moving, fish rippling the water in small moonlit circles.

I have come outside to sit by the riverbank as I write. A cool autumn morning, breezy; my hair blowing. My eyes cannot hold all that is before me, but I sit, I watch. This comes back to me often, the vision of the Indian warriors, the spirits springing from the earth. It does not fade as a dream. I see them clearly. Their faces. Their bodies. Their skin. I hear them. They will not be banished. I love them, their strength, their mystery, their beauty. I feel Webster in my heart.

November 13, 2003 *Blue sky*
Tomorrow is Little Eagle's birthday. His shining spirit fills my heart; his Web smile and dark eyes. I don't know why I am brimming over with sadness tonight. I know it will pass into a sky so blue. The yellow moonlight shimmers across the black river. The trees creak in the night wind.

2004 ~ Mysteries

January 16, 2004 *My own winter*
All I want to do is write. To walk alone along the riverbank, the woods and ravines, climbing, pushing. Finding incredible silent places of beauty like my own dreamy winter where inside things are growing.

January 17, 2004 *I begin my country life.*
Swimming back to myself, tides moving. Breathing. Leaving others behind—solitary. From the riverbank, three herons take flight, glide over water, call their harsh call, so close I could have asked them to take me with them.

I was born into this world a spirit bright as sunlight sparkling, as solitary as the narrow wooded path, as mysterious as night falling on the river.

As a child, in my house there were rules about how to do things, how not to do things, and I was closely watched, kept safe. There was great cleanliness and order. I loved the world of forests, building forts, swinging on vines.

When I graduated from a college near my home, I lived in an apartment for a year, close to my parent's home. But in 1976, I began my life in the country. My house was a small log cabin. There was a big, gleaming black snake that lived in the twisted tree out back. I looked for him every day. I was newly married to Warren, my adventuresome husband.

I began my life of wilderness. We lived on the Chickahominy Lake that flowed into the Chickahominy River. We wandered along that quiet river and my heart opened.

Our house was small and crooked in places, with two closets. We put up bamboo curtains, painted the bathroom and kitchen green. My kitchen was long and narrow. The windows were drafty. I loved our small log cabin. My neighbors were my husband's Aunt Edna and Uncle Irvine. I love them dearly now as I did then. They are kind and loving, the best this life has to offer.

Who were we? We were young with our smooth, glowing skin, our thick, dark hair, our shining eyes, our hopes before us. I taught English and French to middle schoolers. I struggled the first two years, but I loved my students, and they loved me. I saw their hearts, their hopes. There was room in that classroom for curiosity and silliness.

January 18, 2004 *Swimming back*
Rainy day. I wonder what is taking place beneath the surface of the river. I imagine all kinds of fish and snapping turtles and snakes gliding into the brown water. What else? What else?

We come into this world with our true nature, then life comes at us, and we layer ourselves. I think we always yearn for that naked self, and we keep swimming back to that place.

January 19, 2004 *The breath of winter*
Sitting on the front stoop, bundled in hat, scarf, long johns. I walked in the sunny crisp cold. Many seagulls flying today, white, white against the sky, the water. I wanted to stay bundled against the cold, to feel the breath of winter on my cheeks, in my lungs, wrapped in solitude. The wind is gathering and moving across the land, the trees, the river.

I want to look at my Linda McCartney cookbook. I don't trust store meats with the antibiotics and hormones, and they don't taste good. I like the catfish from the river or deer. I am thinking about a garden this spring. The reservation gardens are so pretty.

January 20, 2004 *High stakes for this river*
The morning before work—watching the river, the sky as I move about, making coffee, doing yoga stretches, dressing. Wanting to hold the river inside of me all day.

At my office in town, needing to recall the early morning at home, the cold river and sunrise. Getting ready, I paused several times to look out, to look at Web's photo, to ask that the Virginia Marine Resource Commission (VMRC) vote to appeal the judges order to hear Newport News again. The VMRC already voted unanimously to deny another hearing. Like a taut game of checkers. High stakes.

I wanted to sit quietly by the window and write. Now, the tedious tasks of work have my mind drawing images of the river…

January 21, 2004 *I think of tiny seeds.*
I am weary of people who speak empty words, oh so weary, so I make myself swim away and unto myself. Always. Feel the Chief's love, know he understands what it is to call up your strength. In the cold of winter, bundled in hat, sweater, jacket, scarf and gloves, I think of tiny seeds to plant for my spring garden.

January 29, 2004 *Glittering sunlight and snow*
Quiet except for hundreds of gulls flying all about and landing on the river. Now the sun is opening the day.

Headed out for a walk before work into that glittering sunlight and snow. As I got to the passage of spirits, I thought of something from *Bury My Heart at Wounded Knee*, "And they lay in the snow and died." The big kill off. I sat in the snow and thought maybe they died but they did not really leave. I saw the life slipping from their bodies and saw their spirits journeying to another world.

January 30, 2004 *And that is surely enough*
Maybe what I love about this river, the earth itself with its moss, the amazing, whimsical fiddlehead ferns, the sudden sight of a still, silent heron, or the wind and wings above me—a bald eagle… maybe what I love too is the mystery in myself that amazes, that is still and silent, that takes flight.

I have been missing my children, my daughter now twenty-three getting ready to move to Ohio, my son, nearly nineteen, not home much. I see them before me. My newborn daughter who looked like a china doll with black hair—a little "Pegapoose," we called her. I can smell her still, holding the musky scent of another world, pure and sweet. I cannot stop placing

my lips on her head, cannot stop staring into her
eyes, her soul. I nurse her, and we are one heart it
seems. I see her at three months in the stroller, dark
eyes mesmerized by the light playing in the trees. A
November baby, I protect her from the cold, bundle
her in soft zip-up jammies. She always wants mama. A
toddler, walking, we spend much time outdoors, visit
the cows across the road. The month of March, we sit
by a small cow pond, winds dancing across the water.

My son, gentle from the start with his dark hair that
would later turn light. The nurses in the hospital
nursery told me he was sweet and undemanding. I
kept him close in the bed and smelled his skin. He
was eager from the start *to be*. He wanted to explore
before he could talk, before he could walk, pushing
forward, pushing beyond the baby gate. Let me out
into the world, he was saying. Strapped on my bike
in a baby seat, "Fla fla fla," he said over and over,
pointing along the bumpy gravel driveway of our
old home surrounded by fields and wildflowers. I
stopped the bike. I looked at him and listened. He
was pointing to a wildflower on the side of the road—
"Fla"... Flower! He wanted a flower! He knew what
he wanted and kept seeking it. I listened. I picked the
flower and gave it to him. He clutched it in his hand
and was content as we rode along. Now, I listen still as
he explores what he wants to do about school, work,
life. Still he sees what he wants and keeps after it until

he gets it. He is still sweet and gentle, and he always makes me laugh.

Both of my children see to the heart of people right away.

Sometimes I think of all the things that make me strong, things I do well, and many are quiet things people do not see directly, but I think they sense an energy in me. A friend told me that when he met me, he knew I had a strong karma. Sometimes it is hard to have these quiet things only or nearly only. But as I write, I think I am like the river, the moon, the forests, and that is surely enough.

February 5, 2004 *Reservoir battle*
Last night was the first subcommittee hearing for House Bill 797 granting permanent easement into the Mattaponi. They "passed it by" which equals voting against it. Next week it goes to committee. Today is Senate Bill 420 at 1:00, subcommittee also. I am not relaxed. It is so hard to sit and listen to their lies and think of the outcome of their lies and what they are trying to do with their money. But so far, we are on top. Carl spoke well but is weary. Over and over, he stands tall and speaks these words. There are hundreds of small stones in my stomach. Slept poorly. It is the

whole situation that hurts and angers and leaves you hoping and praying.

In the miserable, sleepless night, I kept calling up the river flowing, waters of the Mattaponi, the spirits of my vision and always dear Chief, your love. So Web, for now, the river is safe. But as you said, "I am a man and must endure what other men are doing." With strong heart I face this day and hope with all hope that in the end (and soon) we, you will have endured their years of schemes, that the river will remain safe, your people will be safe.

Trying simply to breathe. I think of Hurricane Isabel in September, winds gathering, gray, wild skies, trees bending, the river filled with waves and whitecaps, the air with its eerie voice, an eerie scent, and yet beneath it all, the river flowed on.

February 8, 2004 *To be a human being*
Finished cleaning, laundry, phone calls, reservoir tasks. The sun is bright, the sky blue and cloudless. The river flows, and the wind roughens its waters. Tomorrow is another day of battle at the General Assembly. I am gathering my strength, wanting to keep light within. We lost round two last Thursday. Much corruption, and it is hard to bear when this river and the Tribe are at stake. Carl spoke with

dignity from his heart. Afterwards, Senator Hanger leaned back in his chair, tapped his pencil, feigned casual interest, and said to Carl, "I'm just curious, how much Indian blood do you have to have to be an Indian?" I wonder how much blood it takes to be a human being.

Going outside to work and feel the sun and wind.

February 11, 2004 *We will save this river.*
The House tabled House Bill 797, which essentially means they killed it because there is no more time for them to meet again before the bill crossover—House to Senate/ Senate to House. Senate Bill 420 goes to the Senate Agriculture Committee Monday for a vote. They delayed the vote one week because Newport News knew they did not have the votes to win. Newport News immediately hired three more lobbyists at more than one thousand dollars per day, in addition to the four lobbyists they already have. Tabling the vote only helped Newport News not to lose this vote.

The night before, I slept peacefully, oddly peacefully. It just came to me that it was almost like the night Web came to me after he died, the peaceful quality. I was asleep; I did not wake to his presence. I did have a dream though, a vivid one. Carl looked at me and said it was going to be fine. The peacefulness has stayed with me, and I think we will save this river.

February 15, 2004 *The goldfinch flew into a nearby tree and washed the stone from my heart.*
Cold. White skies, the river a silvery gray. In this day and time, to have this pristine river, and I am here to behold her every day. That someone is trying to destroy it all makes it all the more precious. I had a restless night. I kept waking, trying to shove away thoughts of Newport News. What struck me was the audacity and the arrogance that Newport News wants to destroy an entire ecosystem, and there are other excellent options for water. It is only that they want it. To own. To sell. To control. They want to turn this river into a manmade nightmare of development.

The Mattaponi Indians were essentially forced onto this reservation, and now Newport News wants to take what little is left of that way of life, to control, to dictate, to surely offer the insult of money. What is *money?* Ahh, but what is this river? It is unthinkable that it has come this far. And yet, look at history. Look at the earth to see what has always taken place. The truth of what was done to the Indians comes blindingly before me, takes my breath away.

February 25, 2004 *A lover and a warrior*
Every day I wake up a lover and a warrior. I look out

at this ancient river and am filled with its beauty. Every day I brace to fight for it. Every day the Chief is with me, Webster Little Eagle. Every day the spirits who once lived here are with me. Every day I pray for strength, for love, for the river. Every day I pray that Carl may realize his vision of the river flowing as it has for thousands of years…for *seven generations to come.*

March 5, 2004 *Leaving behind my whirling confusion*
Warm and windy, blue skies, puffy clouds, thin loose clouds, current running fast, marsh grasses bleached white, deep blue river, trees brown and naked. I have left behind the doors, the walls, the things we accumulate to live in this world. I come to leave behind my whirling confusion. My mind is a restless blur. I am, dear Web, "out of kilter." Pen aside, thoughts aside, I desire just to be.

March 21 2004 *I dreamed last night of your face.*
Sunny windy first day of spring. One year ago since you left this world, Webster. I have surely missed you, but you are in my heart, as I breathe, as I look out upon this world. I dreamed last night of your face; you were close, then you were gone.

March 23, 2004 *Choosing love*
Woke to sunlight, mist on the river. Shad nets, fishermen. I have so much. Wanted to follow an internal wilderness, a solitude this morning, but I had to leave for the day. Looking back on my life, I think about choices I made to fill needs and wants as best I knew, and I understand somewhat. I have not wanted to make demands on someone's heart, be it a friendship or a love relationship. Either someone wants to give love or they don't. I believe that when you give, you give because it is what is in your heart, and there are no strings, no conditions, no trade-offs. Maybe I have left myself open to hurt because of this. Yet, I know that when I make a decision, when I act, I must choose love, not fear. This applies to my writing as well. I cannot allow anyone or anything to keep me from this.

April 5 2004 *Shine through me.*
Full moon tonight. Wild wind, wild river, whitecaps, clear blue cloudless skies. Waiting for Robert Pruett and John Henley with *Pleasant Living* magazine to arrive to do photographs for my piece, "The Chief and I: A Journal" and the update on the reservoir. Cleaned the house, did yoga, primped, and cleaned the glass on Web's photo. I guess we are ready Web. Fill me with your love, and shine through me.

April 16, 2004 *Mirth*
Alone on the pier. Sunshine, windy, blue, cloudless skies, Cool but still in my bikini. Sun on skin. Good to have nothing but this river.

One rainy evening this week, Tuesday maybe, I took a walk after a long day of sitting at my desk in town. It felt so good to be out even in the grayness, for some reason especially in the grayness, the air in my lungs. It was getting dark, and I was walking briskly to get back, walking in the passage of spirits, when suddenly I stopped, when I felt as though something stopped me, asked me to pause. Oh, I had not lingered there for days, and I felt mirthful! Felt the spirits had missed me.

June 11, 2004 *My wilderness*
I record the dates and days as if the measurements matter, but I wonder. I know the sun is out, the hot breeze blows across my skin, the river flows into forever. I am in *this* world right now. Did I come from another place, another world? Do I slip into other places? These words of my wilderness that no one touches.

June 14, 2004 *Becoming*
Silky river and reflection of white sky, green marsh.

Do I have the words to say what is in my spirit? This morning, early, I came out to the quiet beginning of day, the sun a soft yellow in the sky, deep blues pale blues soft soft, and did I say how quiet it was, the river gentle. And now another quiet end of the day. I peer up from my journal, and the yellow sunlight fills the clouds. Ahh… but the morning I swam, the water velvet and cool on my skin, and I was filled with the beauty until I became the beauty, and my eyes all day swam and held and sang. In my fingers, the mystery seems to be alive. I feel as though I am becoming more the spirit of this river and sky and land.

Last night the fireflies from my bedroom window lighting up the woods, the ravines.

June 16, 2004 *River heart*
In the early morning, before my mind has woken up, I make my way to the pier with coffee and look out on the opening day. I stand before the brown water. I dive into the silent, dark world and feel my body surging upward, hair wet and gleaming, skin wet and gleaming, and I smile and begin to move through the silky river. The fish splash. We are together in this riverworld.

Every day I swim and swim as if I am flying off into blueness. I drive to work with river hair and river skin and river heart.

July 5, 2004 *Tree frog*
I sit in the winding ravine, the soft embrace of trees and ferns. Big green tree frog greeted me this morning as I walked out into the morning. Hopping with its strong, cheerful body.

July 11, 2004 *About gentleness, about strength*
Dark blue butterflies with turquoise stripes flutter and light on the white blooms along the river's edge. Dragonflies all about this year—red, gold, blue, green, and even turquoise. I have come up from swimming in the river, cool brown water, sandy bottom, muddy bottom—a place I trust. Emerging from a place that tried to keep me from love, a place that left me with a heavy heart. Emerging with a smile. I am learning about letting go of obstacles to love, about gentleness, about strength. Oh I am becoming as brown as a berry in this riverworld, sitting with the sun, writing. I think I always held and held my spirit no matter what pushed against it, until I became as free as a butterfly. An instinct always directed me to nature. As a young adult leaving my childhood home, it was no accident I am sure that I chose a rural life, a wilderness.

July 12, 2004... *over and over to soar*
Indians. A book shows a photograph of Indian children. Taken away from their homes. Hair shorn to look like whites. Clothes like whites. Their language forbidden. Their ways forbidden. The shine is lost in their eyes, the laughter gone. A sorrow too deep for words. I believe that beyond the photographs, they tried over and over to soar, but I think the reality is that they lived in despair. I think of Carl in this white man's world, told in his youth that the Indian ways were wrong, taught by white teachers on his own reservation that his people were savages. He has held to his inner strength and kept a kind heart.

Why do we keep striving? Like the shad that return to this river each spring, we seek love over and over.

July 25, 2004 *Such adventures*
The rain is falling softly. I sit beneath the canopy of leafy trees. The wind on my skin, the brown river moving, evoking what mysteries are inside of me. I have had such adventures since I last wrote in my journal. Last Saturday, I took the longest swim so far, to Trimmer's pier. The morning was full of sun, and I just kept swimming, watching the green banks, a kingfisher darting from branch to water, skimming across the surface for fish, then back to its perch. The water is cool and glossy on my skin. Gliding

along, I feel a part of all I see and touch and hear.
At Trimmer's pier, a simple finger out to the water,
I make my way to the beach, his steps, his yard,
his vacant fishing house. I respect its modesty, how
it blends with its river place. Webster's marshland
borders his, and I summon my dear friend, think
of his footsteps, right where I am. Mr. Trimmer
raised bees, I see; bee drawers sit on a high hill. My
imagination makes me smile and wonder. He lives
down the road now and used to bring his children to
fish and swim at this house, from what I heard. No
electricity, just the river.

All week I swam, mornings before work, on my days
off, and one evening when I had been in town until
7:30, getting home at 8:30, I knew I had to swim,
as light was leaving the day. A yellow crescent moon.
The white flowered bushes ramble all along the banks,
the wind carries their scent across the water as I swim.
The tall, red flowers along the river's edge, I discover
them one morning—Cardinal flower. A great blue
heron takes flight.

July 30, 2004 *The only power I have*
Days of wild rain and storms. Now today the sky
seems bluer than ever before. Swimming past my achy
body on and on, I realize that I don't want to be right
or to control someone else. I don't want to be better

than. I want only to be as good as the person I am, and it is a freeing thought.

We make it through darkness, and there is no way around it. Each day we face it, then somehow, light seeps in and we walk into it and keep walking. And it comes to me that the only power I have is love.

July 31, 2004 *Full Moon*
Two full moons in one month ~~~~~ "once in a blue moon." Everything in its season…

August 13, 2004 *Reservoir battle, again*
We lost yesterday after two days of hearings before the Virginia Marine Resources Commission. Last year, they refused the permit, but this year five of eight VMRC members sold out. Politics and money. If you cannot win by the rules, break them and lie. Carl was quiet. I am overwhelmed with sadness and anger and do not know how to bear these feelings. They ramble around my heart like restless ghosts. I see how life can chip away at your spirit, but you keep reaching for light. Today, I am ready to stand up for the river again.

August 28, 2004 *That I be exactly who I am*
Home after being with people for several hours. Quieting my mind. Nighttime has become a restless time of keeping the demons at bay, the reality that threatens this river, wondering what is worth our breath if all that is real and beautiful is not safe. I keep reaching for joy.

During the day at work, how can I pretend a world that is balanced? I swim from anger to sadness to emptiness to warrior and protector of this river to lover. It seems to drive all that I am.

And I come again and again to the water to find peace, silence from the outside world, from my restless mind. When I am in the world, it feels essential that I be exactly who I am, speak from my heart, look people in the eye.

August 30, 2004 *A wakeful dream*
A lover and a warrior, every day, every breath. Walked into the early sunlight, bare feet on the earth and moss. Wind in my hair. Trees folding over me. Yellow light washing between clouds, slashed on the river, moving gently, my heart widening, my heart reaching, my eyes holding, my breath praying.

Afternoon, work in town
A dark storm is gathering, lightening, thunder. A full moon tonight. I have been feeling as if inside of me a dark storm is gathering that I cannot keep at bay. Yet… as I worked, what came to me was a sense of the warrior spirits from the river, as if I were with them, as if I were in a wakeful dream, and they held me within their circle of bodies. They touched my face with their fingers oh so strong and oh so gentle and Webster touched my head. I felt peace and love.

August 31, 2004 *Protected*
Yesterday on a full moon rising, a sudden, violent rain storm, Gaston, swept through the area, flooding downtown and outlying areas. People were stranded all over. Others drove home on flooded highways. It took me two hours to drive the normal one hour home, and I drove through high waters and thought I might get swept off the road several times. But I was safe.

Now I know why I felt the protection of the spirits, felt Webster's hand on my head.

September 7, 2004 *Defining myself*
I am thinking of the early morning quiet, standing on the high bank with the new light in the sky. The moss

beneath my bare feet. Swimming in the chilly, stormy river, a body moving, moving, spirit rising to the tall banks. I am as strong as the silence around me. And as I drive the rural roads, easing my way into town, to the busyness of work, I think how in my life, I have defined myself in one way, against friends, lovers, children, parents, and those I cannot comprehend. I listen and watch and take it all in. But in another way, I leave them all behind… and define the truth of myself by wildflowers, the morning light, a heron.

River Dusk, the Mattaponi

I went out to the pier and watched
the ending of the day,
the sky wider than my dreams;
the clouds dark as plums.
A deep rose slipped across the horizon
above the trees
above the river.
I stayed and stayed and did not think sad thoughts,
I did not think of loss or fear.
I stayed until all that was left of the light was
a jagged strip of yellow layered across the horizon.
And a wind came along and ruffled the water
and the silhouettes of geese honked.

September 26, 2004 Shall I be too?
Autumn enters the earth in silence. The deep reds across the marsh. The sky now crisp blues and sharp white clouds. The air is cool at night, in the morning. Web has been laughing it seems. I pass his photo, and it is as if he is smiling at me, being playful. Do you like autumn too, Web? Are you laughing because I have auburn hair? Yes, Web, my black hair is no more. But my spirit is the same, sweet Web. I feel the twinkle in your eyes. We are autumn birthdays, Web.

It is the purest joy to be here. The geese are honking; the breeze is stirring. This morning, I walked to get the newspaper at the end of the road; I wanted to see the autumn wildflowers. I love their reckless beauty. Nature is bold in her colours, her designs, her loveliness. Shall I be too?

October 2, 2004 Turkey girls out for a saunter
Woke to the sound of some bird. My curiosity got me out of bed. A turkey! Trotting by the river bank, then another. Two turkey girls out for a saunter, checking out autumn with me joining in.

This gentle, quiet river with autumn filling her banks. All I see—yellow flowers across the marsh, yellow reflection in the water. Splashes of scarlet surprising again this year. Rain showers briefly, now blue patches in the sky. A lime green tree frog has resided just outside my bedroom door for two days now, all sleepy. I gently touched its back this morning. Soft and sticky. As I sat on the pier there was quiet, then turkeys gobbling in the field. They flew out—two, four, six, and eight across the river to the marsh.

Yesterday, two eagles soared above me as I swam. The spirits whisper to me. We must save this river in this crazy world. And I pray for love amidst the wars and hatred and fears.

October 10, 2004 *Returning to the land*
I feel autumn in my body. I think the swimming is over for this year, except for my birthday swim. I swam twice last week before work. An eagle soared with sunlight shining on its feathers. And in the evening, the eagle soared again with sunlight lighting its feathers. It was cold swimming. I knew somewhere inside of me that the cycle of autumn was coming, and I was drawn to be a part of the river again and again. I seem satisfied now to return to the land, to walk through the autumn woods, see again the thick moss, the ferns, the wildflowers. I feel peace only here, shadows and light playing on the river. And I have said that Webster seems close, his spirit laughing and joyful. I keep asking him if he is coming back in his grandchild this spring. We will rejoice in the birth, the

river flowing *for seven generations to come.*

October 14, 2004 *Blue Jay squawking*
Time is moving forward as I sit and sit at the computer at my job in town. Unable to stay at my desk any longer, I walk outside into autumn and find a tree on fire with orange leaves. I gather some leaves, "For you Web," I say, my soul opening. I hear a blue jay squawking, see him hopping from branch to branch, the loud bird. "Write your journal," he squawks to me.

October 16, 2004, *Gifts to mark us, to make us strong*
My father. I recall walks with him in the autumn woods. I loved our adventures, tromping over leaves, branches, two ramblers we were. He read poetry to me, taught me to dance to Glenn Miller, could never get enough hugs and kisses. And in college when I was struggling with student teaching, he made me a roast beef sandwich (with lettuce on both sides!) and brought it to my dorm with a fat piece of homemade chocolate cake my mother made. He reminded me that I was strong, that I had much to offer. My father, the only child of poor immigrants, made a good life for us through his integrity and hard work.

My mother. Beautiful soft skin and dark hair. Her love, always gentle. I see myself with my mother and sister, little girls, wide-eyed, walking to a nearby pond to feed the ducks. When I was in high school, my mother let me stay home at times so I could sit outside and write. She understood. She did not judge. She is the best cook I know, famous among my friends for her delicious Armenian foods and her Armenian cookbook. My mother has the spunk to find a way to do what is important to her.

I am the granddaughter of survivors of the 1915 Armenian genocide by the Ottoman Turks, a thread that is woven into who I am.

In my family, I knew I was loved. I knew I was safe. Simple and deep. I had all of these gifts to mark me, to take out into the world, to stand strong.

October 25, 2004 *My dreams and visions*
I am amazed at the intensity of my dreams, cannot recall such a time in my life. I have considered them, tried to find out what they represent. I have some ideas, and now I want to let them go. They haunt me, so I am hoping that if I "hear" what they may be saying, I can release them and feel more at peace.

My autumn walk Saturday into the deep woods, the steep winding ravines until you look up and suddenly you are a small creature among the towering banks, the tallest trees reaching to heaven. Walking toward the reservation, I am a small breath in this place, but, indeed, a breath. On my return walk, the sky filled with hundreds, thousands of blackbirds, their wings in flight, woosh woosh, a sound to behold framed in the silence of the forest. Their patterns of flight amaze. Could not help but feel the spirits of hundreds who came before me.

I recall the dream of the owls as my father-in-law was dying in 1994. George was near the end of his battle with cancer when I dreamed of three owls. They appeared to me in a setting of a lush green forest, and the owls were an incredible salmon color. I had never seen such intense colors and beauty. They were gently tugging at George, reassuring me that he would be fine, and that they would take care of him. I feel they were in fact angels, and what I saw was the beauty of heaven.

I think of the vision of the warriors, of angels and spirits.

What does it mean that I have such dreams and visions? That in my dreams, I see the smallest details. I hear. I smell. What purpose might they serve?

October 26, 2004 *Curiosity*
More and more, I am impatient with "experts" who rely mostly on being book-smart. It seems like a barrier between the self and nature. Where is the common sense, the heart sense, the gut? I want real body and mind, real heart and spirit. Knowledge is within each of us, and if we don't first search the self for answers, the knowledge we gain from education, books, or others has little true meaning. It is regurgitation, words filled with air. I think we learn based on a natural curiosity, and then, and then… we follow this curiosity for answers. We enter the world full of wonder, and too often it is dulled over because it takes too much time and energy to nurture.

October 31, 2004 *The thing within that makes me shine*
Golden leaves cover the ground, the steps to the pier. My feet brush through them, stirring the smell of autumn. Soon the air will be cold, and the winds will blow the leaves off the trees and we will make way into quiet winter. My birthday is Thursday, and I feel changes inside of my body. Part of me resists, but what good does it do? I tell myself that I am like autumn, and I will move into my own winter in time. I will make preparations somewhere inside of myself. Colors fade, but there is still the beauty of winter. Wisdom deepens in old age, an innocence returns. Do we let go of worldliness? I think of Webster who

never seemed old to me. Surely he had troubles from his aging body, but his smile and eyes never seemed old. When I look in the mirror as I age, I have to see beyond the image. I have to call up the thing within that makes me shine.

November 4, 2004 *My birthday – Another trip around the sun*
I felt special even though the weather was rainy and gray. Before the heavy rains descended from the skies, I walked before work to the rivertop of the world and tromped around in the woods. Then I made my way to the river and jumped in three times. Once for me, once for the river, once for Web.

November 13, 2004 *Mere survival*
I have come to understand the ill ease I feel in the world. It is not terrorism or a direct fear. I don't feel threatened. It is the energy of being out in a world where there is too much, too many. Buildings spreading on and on and an endless sea of cars. Too many people rushing about, too much "purpose" that has no purpose. Too little love and kindness. Mere survival.

What will happen when the secret corners of the forest are heavy with the scent of too many men? When the hills are blotted with too many talking wires? It will be the end of living, and the beginning of survival.
—*Chief Seattle, Suqwamish and Duwamish*

November 15, 2004 *Jessica's birthday - Haunting dreams*
Father Hovsep, our Armenian priest, and his wife came for dinner, and Father Hovsep blessed our house. He blessed us, the river, the land and woods. And Webster in his photograph.

The dream: a small, soft animal, fawn-colored at water's edge in a forest, at first cuddled to my chest, but then I could not pull it off. I called for help to a doctor I know who was in the woods hunting just behind me, but he did not respond. He operated on Jessica's finger when she was five, and his office is directly above mine in town. I pulled the animal off of me, and I was left with a wound. An owl appeared and did not leave as I got closer. It was a small, bluish gray owl who sat very still and looked at me. I woke filled with the haunting of this dream, and it has stayed with me for days. …When I got to work the morning of the dream, there was an e-mail from a colleague whose last name is the same as the doctor's. He had sent a photo of his hunting trip—a small, soft animal nestled against a tree.

On my way home from work the evening of this dream, I was almost at our house. I was driving on Wakema Road with woods and deep ravines on each side, and an owl startled me and flew out right above my car, its wide wings lit up by my headlights.

December 17, 2004 *Owls*
So many owl dreams and real owls. I think I
understand some things at some internal level, yet
they haunt me. Why? I heard two owls last week
outside in the yard, right outside my window
that faces the river. They were beautiful sounding,
fascinating, yet I could not help but wonder… more
owls. That night or the night after, I dreamed of
two large spotted owls trapped in the garage, and I
opened the door and let them out. They flew off with
wide wings in such beauty. They seemed so real, their
brownish feathers with white spots. Now I am feeling
Web so near. Is he coming back?

December 20, 2004 *Between two worlds*
Strange days, oh so strange. I feel as if I am between
two worlds.

December 29, 2004 *A long ago life that whispers in my dreams*
The cold bites. The gray world embraces me. I must
be strong, to hold against enemies, to hold against
desire for things to be different. Listen to the small
voice that declares there is light, that whispers *blue sky*, a glimmering river beneath moonlight. There is
always the soft hope that trembles when your fingers
touch the earth. Eyes of another time look out and see

these waters hundreds of years ago, a long ago life that whispers in my dreams, recalls cool water on skin, his dancing eyes, his breath is yours.

December 31, 2004 *Peace and love*
Looking out on a thick gray fog—I cannot see the river but I know it is there. Sitting at the dining room table with coffee and candles. The ending of a year. December, the month of popping trees.
My time has been rushed and scattered, and it has been too long since I sat in the quiet morning.
I found this place to live my life, or maybe it found me. But nothing is safe. The dogs of hell want this riverworld too. It is easy to lose hope, and hard to find joy, but I have no choice or my spirit withers.

I am fifty-two years old, and my body is changing, its rhythms moving. I walk each day, balancing my spirit that feels timeless with the ticking of this world. It is a world I cannot help but think is destroying itself. Still, still, I live in it with a loving heart as best I can. Webster has become even more a part of me, and I am not surprised. These past months have been filled with dreams. Owls visiting me in my sleep and in my waking. I am not going to analyze, but I do have a sense of their place in my life. I have described my dreams, my visions, the two owls in the yard, the one that flew in front of my car. I recall that a couple of weeks ago, I sensed a change in myself, and that the next day, it became stronger. I was filled with a beauty. I was filled with this and only this. Each day I walked and was filled too with the sacredness of this riverworld.

On Saturday, I took a long walk to the rivertop of the world where I can see far across the river, the horizon, the reservation. I felt a powerful and deep connection to Webster and asked that he grant me peace, said I would have his spirit live through my love. I had a strong sense that this is what he was asking of me. The night that he came to me after he left this world was once again sharp and clear.

All of these days, I thought of my dreams, the vision, the owls I heard and saw. On Monday, I woke with a peacefulness. Even my body felt light and energized. It was as if on this day, the two spotted owls in my dream, the ones that were trapped in the garage, the ones I freed that flew off in such beauty—like angels—became clearer to me. My spirit stronger, Web's spirit at peace within me.

With this peace and love, I begin a new year in a world that has too little peace, too little love. Though sometimes it seems as if I do not know much of anything, I will try to be kind and open and listen. So many stars tonight.

2005 ~ Riversong

January 5, 2005 *For the sake of flying*
I step out into the early morning, my feet finding soil and frosty field grasses, leaving behind house and road, entering a wild place. The darkness enfolds me, then light begins to enter the day. I look up and see black wings, white tail feathers, the yellow beak. The eagle flies above me, circles, his life gracing mine. There is freedom in knowing we do not have all the answers, and we fly for the sake of flying.

January 13, 2005 *The moon hanging just so*
Afternoon, work in town
The trees are swaying, the leaves swirling through the air. The sky one moment dark, the next a bright blue bursting through. Like me I think. These days and weeks I have been a turbulent river, a quiet river, flowing despite all around me, despite all within me.

Home in time to catch the last light of day. I went to the pier and sat in the quiet, by the river, open and wide. The sky a soft steel blue with washes of white. I sat with knees drawn to my chest, and a light caught my eye. I turned, and behind me toward the shore was the crescent moon reflected in the river and the black trees and white washes of the sky reflected in the river. The moon, wavy in its watery world. The moon hanging just so in the sky.

January 16, 2005 *The geese are calling*
Bundled against the cold, sitting in a deep ravine, the river flowing beside me, wind, snow flicking lightly on my face. Bundled against a place of people who seem to set a high value on being "right." Bumping up against those who seek you because they see your light, and it gives them hope. But you cannot sustain it for them, and soon like a shadow, it darkens the light. The geese are calling.

January 23, 2005 *Love endures*
A glittering sunrise on icy trees. All that I am, all that I know.

Funny, people tell me I am good with words. I hear others use words in ways that seem like a game, for control, to hurt, to put up walls, to anger. Sometimes they withhold words to withhold themselves. I think sometimes people are afraid and afraid to love, and I don't think they know it really, or they don't know why.

It takes so much energy not to engage with people who act out of a place that does not come from the heart. For me, it always comes back to love, and when that is real, your heart knows it. You can look someone in the eyes; you can hear their words, even if the words are not always easy to hear, even if they mess up. If the love is strong and true, it endures.

February 27, 2005 *Baking Bread*
Flicker of candles, winter river before me pulling me toward her, keeping me alive. Walked in the woods into ravines to the rivertop of the world. Can see the winding river forever, the reservation, pausing against a tree to watch, to breathe, to know. Everything I touch, smell, see, becomes a part of me, or I become a part of—the crooked tree, the owl who-who-whooing in the night, skin, hair, the fire from the candle, my babies, their smell of sweet musk, memory behind closed eyes.

I have been preparing bread since morning, mixing, rising, punching down, kneading, shaping, rising once more, preparing an oven with water at the bottom in a pan, brushing the loaves with salt water. It is called "pain ordinaire" which in French means ordinary bread. It is a simple, plain bread to prepare. It frames my day. At day's end, the smell, the taste of warm bread and butter.

How do you make bread, people ask me. I think the way people make bread is like their personality, a rhythm, a touch, having a sense of things.

March 20, 2005 *To the tips of my hair*
First full day spring tomorrow—the second year of Web's leaving this world. I walked on the beach for a long time this afternoon, a bit warmer and some sunshine. A storm began to darken the sky upriver into deep blue. I thought, first day of spring, Web's passing, needing courage, the river the river… fighting for her. I sat on the steps and took off my shoes and socks, my blue jeans, my shirt, my earrings and tested the water. Cold cold of course. I stood and looked out, then slipped in to the tips of my hair.

April 2, 2005 *Singing across fields*
Walking in a spring rain, lifting my face, singing across fields, the river swollen and shining. As I near the end of my walk, I come to the fallen log at the shoreline and sit as quietly as the river. My hands dip into her waters and wash my face with her secrets. A balm to my restless mind that struggles to be in this world. I am beyond doubting myself. I know I have gifts to offer. The face I wear to others—friend, helper, listener, protector of the earth, bread maker. I know that others see a shine, a passion in me, but

how to earn my livelihood in a meaningful way. Keep planting seeds I tell myself.

April 6, 2005

River Goddess

Birdsong trills through the trees.
Geese honk across the brown river
all soft ripples and sunlight.
The shad fishermen's voices carry through blue skies,
syllables steady, now rising, now falling away.
Like a river goddess, I sit in the hot sun.

The moss reigns prince along the banks.
The pink buds throw off winter.
Somewhere, an artist washes colour on paper
creates form and shape,
while I, a solitary river goddess
for one breath
for one afternoon
capture all I need to know
and all I can never say.

April 10, 2005 *Shad*
The white shad bush has bloomed along the river. The shad fish are running in the river now. The connections must not be broken. This river must not be transformed.

April 23, 2005 *How she wears the earth*
I begin a new story in a new journal. Who knows what it will tell? Laughing out loud as the breeze moves against my hair and skin, as a glossy river gently flows, as the easy sun glistens on the water. Laughing because the weather experts said there would be storms. Geese call as they land in the marsh, their wingspan wide. A turkey in the field behind me makes a wild noise, a mating call.

Ah! Webster's great grandson was born yesterday, April 22, 2005. 7 lbs. 11 oz. Carl's grandson—his first. It makes him smile. Connor Eli Custalow is born into this riverworld.

Few shad this year, fewer than any other year, says one fisherman. I always wonder what unseen reasons may be at hand. Man's hard hand. Five geese fly above me, and I hear their wings ~ woosh, feathers against air; only a moment. Fish jumping. Water splashes. In the end, are we only a moment? Silly questions… that do not matter. I want to be a gentle hand on this earth with a strong spirit. We live in a time of much evil. Too many of those in power are wolves who don sheep's clothing, but oh their teeth are sharp and gleaming. Must choose love over and over.

I have resisted spring this year, reluctant to give up the quiet of winter, her refuge, her simplicity. How she wears the earth in stark lines black shapes against white skies, against snow. Swans and seagulls on an icy river. Now the trees have leafed out, and wildflowers and mysterious plants surprise me on my walks. Fiddlehead ferns opening. Everything opening. I walk and take spring in slowly. I like the hot sun on my winter skin.

April 24, 2005 *A smile in his voice*
A fishing boat is coming, their nets out early. They come each day to catch what they can. A beaver swims and dives beneath the surface. Carl saw his grandson yesterday and called on his way home. I asked many questions. What color hair—*blond but not much*. Did you hold him?—*Yes…* a smile in his voice. Did you smell him?—*Yes, because they were changing his diaper*. Did he have a pretty mouth—*yes,* as if he were seeing it in his mind's eye. "I can't wait until they bring him home," he says.

April 25, 2005 *Beside still waters*
Sunlit river. I walked into the late afternoon finding a warm wind, hearing hawks whistling through air, across fields. I left this morning early. Left the quiet sunrise, birdsong opening the day. I never want to leave, never want to leave… and now I am back sitting with this beauty.

Walking along the beach with a low tide. I want to shed voices that tell me what I should want. I think about church. The Armenian church is the most beautiful and sacred to me, with its ancient music, the censer—a vessel suspended by chains—used for burning incense, but it is here on this river that my heart and spirit know God, and I think of Psalm 23, the favorite of many, "The Lord is my Shepard, I shall not want. He maketh me to lie down in green pastures. He leadeth me beside still waters."

We do not want churches because they will teach us to quarrel about God, as Catholics and Protestants do. We do not want to learn about that. We may quarrel with men sometimes about things on this earth. But we never quarrel about God. We do not want to learn that.
—*Chief Joseph, Nez Perce*

April 27, 2005 *To be strong*
Restless night, filled with images of those in power as

devils. They seek to destroy, to gain material wealth, yet speak of morals. I see their tight faces, their uptight haircuts (I wonder in a snide moment, if they all go to the same barber), and hear their 'right' voices, and theirs is a very unloving energy. Beneath all that professed goodness, there is a dark underside.

I walked out into the riverworld, saw the fishermen and their nets, saw the fish splashing, the wild things growing, and when my walk was over, I slipped out of my clothes and into the river. Oh so cold! I wondered if my heart might stop. I needed that to be strong. Part of me feels that fate is at hand, but part of me knows that as long as I walk this earth, I must be a warrior.

I wonder how the new baby is doing. I hope I will be able to watch him grow. I love to think of him, so beautiful and new. I look at Web's photo, and they are all connected—Connor, Webster, this river.

May 8, 2005 *Mother's Day*
Windy windy. I sit on the pier with one hand holding a floppy straw hat on my head and one hand writing. My son and his girlfriend just left. I love that boy! He makes me smile. He has a loving heart. He gave me the most wonderful card. In it he wrote that I always believed in him, was always there for him, that he

could always trust me. I know he can do whatever he wants, can go wherever his passion leads him.

I bought glass wind chimes shaped like a spruce tree with colours of green and a sweet sound. I bought a silver bracelet, a simple, strong design with three bands.

I flew to Cincinnati to be with my Jessica last week. I think back to the hospital when she was born, and I had her wrapped in a waffle blanket nestled next to me—my daughter. I was so proud to be her mother. Now, she is my unfailing friend as well. In Ohio, we shopped each day and ate out, things I don't do much of here. This world moves as it will, but I have my daughter and my son.

May 22, 2005 *The circle*
Sun, moon, eyes, seasons. I read the draft of Linwood Custalow's book yesterday, *The True Story of Pocahontas*, and he wrote about the concept of the "circle." How, for example, we come up from the bottom of the circle after being at a low point in life. It is a Powhatan/Mattaponi way of seeing life. I was taken back four hundred years, yet the thread into today seems never broken in terms of who these peoples are. Despite all that was done to break them, they have remained of dignity, kindness,

love, and humor. I see it in my Mattaponi neighbors. I am not putting them on a pedestal, nor am I am romanticizing the Indian. These are real people with real problems and real flaws, but many have been steadfast in the face of this civilization's folly, and their ways have thrived.

And now, after having read Linwood's words, their ancestors seem more present for me along this river, these lands. They are alive in my mind's eye. He wrote of hiding the children in the ravines to keep them safe from the Englishmen—the ravines that I love, that I walk through, sit in, feel the mysteries of.

June 4, 2005 *Wild ghost bride*
I've made my way through another four days of town. And I return to this world. Always the river flows quietly; the birds sing in the thick green riverbanks. Now another wildflower blooms—small pale blue flowers—must look them up in the wildflower book. Mountain laurel petals begin to fall—on the beach, across the pier, as if trailing some wild ghost bride, on the fat mossy banks, on top of the river.

I wonder—do people choose their hearts, or do they choose their fears by which to live? Kindness, or being right? Openness, or control?

June 11, 2005 *The time of yellow birds*
Awake before dawn. Long swim along the shoreline, across the river "to prove that I'm a woman." Solitary. Blessed river. Kayaked to Mr. Trimmer's finger pier and tiny cabin. Walked about and felt a simplicity, felt that nature reigned. I cried for the yearning I cannot keep away but *must* move beyond. This tug of war in my brain that worries about the earth, that I would like to find at least some peace about. I feel like I am weighted too far in one direction, all lopsided. The prayer came to me, "God grant me the serenity to accept the things I cannot change, to change the things I can, and the wisdom to know the difference."

And! As I paddled away to explore a sandy beach full of flowers, I heard a mad dash of movement through the woods. A big turtle with orange on its shell scrambled down the bank fast and furious and swam into the water. Was I disturbing its world? It made me laugh out loud and leaves me with a big, fat smile.

Afternoon
Oh my! The goldfinches are everywhere, flying back and forth among the greenness and at the feeder, this splash of yellow all about—*The time of yellow birds,* I shall call it.

June 13, 2005 *Garden angels*
My garden this summer is just okay, not like the pretty reservation gardens. They are garden artists! Garden angels! They are beautiful! But my lavender is growing.

June 17, 2005 *Radiance*
I want to wash myself in naked grief and in naked joy. I want to feel them both raw. Grief and sadness bring me to my knees, have me begging to know the light again, fingering over and over the pain that brought me here. I walk in the woods, swim in the river, sleep, keep to myself, accept the wind and sun on my skin—seeking grace. Layer by layer, I shed the sadness and walk into radiance.

June 18, 2005 *Falling back in my mind*
I keep falling back in my mind. I am sitting in the quiet of talk between Webster and me, in his house on the reservation. His strong hands, his dark eyes, his voice that spoke of saving the Mattaponi River. I believe in the mysteries of this riverworld. I believe in the seasons that cycle upon her. Now green, now the sun hot. Yesterday, the black and white of winter. Small birds flutter through the mountain laurel. I believe that people do care for such small beauty, for such immense beauty. I think they believe such

worlds can never go away, that someone will protect them, that nothing can harm them. They are wrong. Yesterday, I finished Robert Kennedy Jr's book *Crimes Against Nature*, a book I had to read, things I am compelled to know. President George W. Bush and his vast "army of soldiers" fool far too many through fear and trickery. Bush calls it a "culture of life." I call it a "culture of lies." They are not about love. I wept for a long time yesterday after I finished the last word of the book. Kudos to Robert Kennedy Jr. and to his father and mother who nurtured him.

Today I am quiet, exhausted I think from what my mind took in. Today I started reading *The Riverkeepers* by Kennedy and Cronin. Today I choose to believe in love, to believe as Web and I said, that we must save this river. Web's great grandson, Carl's first grandson was born nearly one month ago. I want to see him shad fish with his father and grandfather.

Under the Stars

I have to say it straight out,
to save myself.
From a heart that holds too much of this world,
from a loss that nightly wanders the moonlit skies,
    the shifting clouds of midnight.
Oh I can smell the earth
the moss
the roots
the rainwater seeped into the soil.

Like a dog, nose running along the ground,
    finding my way back
to save myself—I said that.

I want to make my way back hundreds of years ago
before Wal-Mart and Google,
before one giant step for mankind.
Before Jackie's blood soaked stockings.

Before Wilbur and Orville Wright had a dream.
Before the unsinkable ship,
    music still playing,
slipped into the black silence.

Before blankets infested with small pox,
before the *Nina,* the *Pinta,* the *Santa Maria.*

Yes, there.

To a place where a peoples laughed around a fire,
bathed in murmuring rivers,
fucked gently under the stars.

Drank a tea of wild hyssop to ease the labor of childbirth.
Spoke a language that sounded like the echo of drums.

Painted their bodies and sang the song of warriors.

And left the Earth as gentle
       as it was
when they came.

June 24, 2005 *Only by this grace*
Wind across water. Goldfinches fly yellow across summer woods. Flesh and bones, muscle and skin. I walk this earth for a few cycles. But my spirit is woven with the light of the moon, the owl that visits my dreams. The wind that touches my face. A wild summer storm is rising, a bruised sky, jagged edges flashing. My eyes watch. I cannot hold it all, but it is enough I tell you to fill me, to slip into this flesh, to climb into these bones. The flight of a small bird; she threads herself into my dark hair. And I am beautiful only by this grace.

I do not want my heart to tremble with the fears of this world, with the fears of other people. I want to take love into my hands. I want love to be the face I wear.

River. Shimmering, endless flow. You are not a noun to be described to others. How can I tell a river spirit to anyone?

July 2, 2005 *I carve a round moon*
The river laps softly. I make my way here past bitterness, too many words, empty words, silence, a smile somewhere. Past highways, roads, strip malls, fumes, and plastic bags thrown out on someone's way to somewhere. Past endless sitting at work before a computer.

I ask, do you want to see this wide glossy river? Does sitting with her haunt you with a voice you silenced inside yourself long ago?

This morning sitting on the porch eating eggs and toast, drinking strong coffee with cream. I hear across the river in the marsh thick with green, two owls in a wild exchange. I sit still and close my eyes. Arousal of that wild place inside of me. The memory of the owls of my winter dreams, the owls of my waking, flying across my night path, the owls calling in the yard.

I know their wild calls in summer's marsh beckon
again—don't abandon us, don't abandon yourself.

Now nighttime, fireflies. Wind painting across the
black sky. I swam this morning across the river and
kept swimming toward a brightness I misplaced. In
the evening, I took the kayak into the setting sun
that lay across the water, a radiance slipping past
my furrowed brow. Into the deep quiet, past the old
forest of steep banks. My arms pulling the paddle
through the brown silky water, my body saying, I
want only to be alive. I find a rivulet into the woods
where the wind moves grasses and trees and whispers
and touches my skin. I paddle along the shore until
I come to the place where the empty white bee boxes
sit nestled in the shaded hillside. The riverbanks are
made of clay, and I step outside the kayak. Into the
clay, with a stick, I carve a round moon and a wavy
river, an owl with wide wings. And I smile thinking of
the tides washing it away.

July 10, 2005 *Balancing the heart and intellect*
Been listening to Joni Mitchell the past few days and
watched a movie about her life. She says if you get rid
of the demons you also get rid of the angels. She talks
about depression being the sand that makes the pearl.
She said she went to live alone amidst nature for a
year, that some might call it a nervous breakdown, but

she viewed it as a shamanic conversion. She spoke of balancing the heart and intellect.

July 14, 2005 *Agnes*
I found Agnes in an issue of the *AARP* magazine that had a feature on older people talking about the secrets of happiness. I loved Agnes, a ninety-one-year old Native American from New Mexico. She had raised age spots, maybe from the sun, all over her full, light red-brown face, her white hair was pulled back, and she wore a beautiful red native jacket with a round silver pin. Her eyes were closed, and she had a smile on her face, not a big fat smile, an easy one like the sun sitting in the sky at the end of the day. She said, "I don't feel my age, I just feel happy." I loved her.

July 16, 2005 *Growing up with his grandfather's love*
So hot and humid, can hardly breathe. Seems as if we had a cold spring that went right into a hot summer. At 8:00 this morning, it was hot. I swam across the glossy river twice, made my way back to the wide beach. I sat in the shallow water in the shade of the high banks, thinking, trying not to think. Taking in the river on my skin, the lush forest, the green in my eyes, the quiet, the spirits… to be only in that moment. Then I kayaked, feeling the pull in my arms, wanting to be strong. I asked to see an eagle, needed

to see. And as I neared the bend upriver toward the shad hatchery, there it was. Black wings, white head and tail, yellow beak—*Spirit Eagle*—he soared. I watched until he disappeared into the thick tops of trees. As I returned, I paused at Trimmer's pier. My carvings—moon, river, owl—were still in the clay. I touched my moon.

Spoke with Carl this afternoon and asked about Connor. He says he is getting so big and is smiling. And I smile thinking of Connor growing up with his grandfather's love.

July 30, 2005 *Reservoir Battle, again and again*
The Army Corps approved the reservoir, another reversal of Army Corps' Colonel Allan B. Carroll's four-year, comprehensive study that recommended denial of the reservoir in 1999 and in 2001. The U.S. Fish and Wildlife and the EPA could appeal. The EPA could veto. We can litigate… again and again. We have three cases before the Virginia Supreme court: Chesapeake Bay Foundation, Southern Environmental Law Center, and the two people who own land at the proposed intake (Warren being one of them); and two cases are with the Mattaponi Tribe. The beat goes on, the river flows.

Carving an eagle and bees
Kayaked to Trimmer's pier—had it in my mind to go back. I wanted to carve an eagle, the eagle I see soaring to its nest by the reservation. Wanted to carve Mr. Trimmer's bee boxes, and I did both. Carved bees flying out. Quite pleased with the whole thing!

Feeling safe
I commented to Carl one day that he must be happy in his beautiful house overlooking the river. He said he really was happy and added as an afterthought that he felt safe in his home. My mind darted to his father, grandfather, great grandfather, and beyond, how being safe was the most important thing. Safe from the English, safe from the prejudiced world they grew up in, safe from the present world that is destroying itself at clip speed, and still taking from the Indians.

August 3, 2005 *A World of Yellow Feathers*
Swimming upriver to a curve in the beach, steep forested banks rising before me, white flowers called Summersweet, butterflies and bees, the wind blowing softly, leaves stirring. I notice them—goldfinches flying all about the white flowers, maybe ten of them. Suddenly this place is transformed to a world of yellow feathers in the quiet of morning green.

August 6, 2005 *Tracks*
Early morning swim and walk. In the sand, I see them as far as I walk and beyond, heron tracks and raccoon tracks. I add my footprints!

August 24, 2005 *One spirit*
Walking into the green forest to the river as a new day opens. Quiet world, quiet me. I don't want to disturb the great blue heron on the beach. All is hushed. Green leaves sway in a soft breeze. No heron this morning, high tide, clean river, swift tide. I walk into her waters, sleek and cool on my skin, push against the tide upriver.

What I see: the waning moon, fading Sweetsummer, the wind moving the greenness on the banks, soft as a whisper. An eagle flies off. I see its white tail feathers; another eagle, two soaring. A great blue heron flies wide across the river. Mist settled on top of the river as far as I can see.
What I hear: only birdsong and water as I move within her, water as it laps against the shore.
What I smell: Summersweet, silvery scent of air and river.

I swim toward the marsh, the mallow flowers, then swim back toward the shoreline. I near the beach at our pier and the orange sun is rising, its long reflection across shimmering water. A small green heron flies right past me, perches on the pier, his spiked head feathers shining in the sun. He sits for a long time as I sit for a long time in the shallow water, my bottom resting on sand, watching, gliding quietly in the water. He flies off. I make my way up the steps through the ravine, the sunlight falls on the bright green moss.

Spirit Eagle, Webster, the spirit that is the land, this spirit that I am. We are one spirit.

September 3, 2005 *Healing this planet*
Hurricane Katrina hit the Gulf Coast, and it seems as if another place in the world has descended into hell. Thousands and thousands stranded, hundreds dying. It seems mostly like poor, African Americans, old, children, the ill. The scenes from television look as if nature broke everything apart, buildings and people's lives and hopes. It looks like an endless trash dump. There is blame for a government that did not respond in time, for Bush who had all of our efforts in a war in Iraq. For steps unanswered by Bush and his friends, that were begged for to protect wetlands and maintain structures. So smug and arrogant they have been to use the world in such unloving and selfish ways. Yet they are just a few in a long history of those who are not good stewards of the land. I think about the fact that it is Nature once again reminding man,

that She is most powerful. Man is far out of bounds, and nature makes "big" men small and weak. They are foolish, to think they can conquer and control. It is a mistake not to respect nature. I watch the footage, and I am overwhelmed by what I see, by the wars, the lies that uphold the wars, the damage man wreaks upon nature, the poverty, the hunger throughout the world, the cruelty man inflicts on one another, the damage that the American government continues to perpetrate upon Native Americans. All of it comes rushing forth as one grief, just like that hurricane. I look at President Bush, and he seems insignificant, just a player on this earth. I am weary, and it all seems too awful to even lay blame. It is as if everything has gone beyond that point, and now my heart wishes only for some healing for this planet.

September 14, 2005 *Supreme Court of Virginia*
We are here to continue the battle to save the Mattaponi River. The walls of the anteroom and court are lined with oil portraits of judges throughout time. I counted sixty-two. They are all white men with austere expressions. I am sitting in the courtroom now, in the corner of a bench. Lawyers file in wearing black and gray suits and dresses, white shirts. They are laden with fat stacks of bound papers, bulky briefcases. Such an air of importance and called-for reverence, and in that we may find grace from their

work and these judges, it is true. That we must follow their rules, dodge the missteps and foul plays of our opponents—this is grim work. From the perspective of timelessness, it is nothing. We will rise or fall, and our spirits remain true, the spirit of the river, its lands and creatures, and those who lived upon and loved these places before us—they will remain grand in the presence of these small acts. I sit here, in the enclosed, windowless, air-conditioned building that will fade and crumble with time, while my mind is on the Mattaponi River. I feel the wind on my face, hear the lapping of the waves.

The judges have entered, one black man, three white men, and three white women. I would guess their ages to be from forty-ish to sixty-ish. The arguments are being made; the clock meting out the river's fate. Reasonable voices, reasonable words; all evade what is true—unnecessary loss, the earth spiraling into destruction. I sit forward and listen, trying to grasp the words, the process of law, while my heart is heavy, my mind angry, a stone in my stomach. Sadness swirls in my being, and I hold back the tears. I want them to stop talking. I want them to stop the reservoir.

September 15, 2005 *Not without hope*
Time has settled my thoughts some, and I am not without hope for the outcome of the Supreme Court

hearing. The court can go against the treaty, which I don't think is likely. How would it seem for the Virginia Supreme Court to say that an Indian treaty is worthless, especially with Jamestown 2007 coming. They can pass on making a decision if I am correct, yet I would think if they cannot make a decision, who can? It seems like that would be lame. They can decide in favor of the treaty, which would end this whole nightmare I believe. I suppose they could say the treaty is valid, and say the reservoir is not violating it. And many other scenarios I cannot fathom. Two months is the guesstimate for their decision… which to the day, is Webster's birthday.

October 1, 2005 *Upside down*
Crisp sunshine. The wind blows, falls away, rises again. In the hammock, I have come outside hoping for grace from a silly sad place. Nearly fifty-three years on this earth, and I am missing the ease of my younger body, though I have more ease in my heart. Mostly I fret over how it will be not to be young, wishing for a crystal ball. My mind says to let it be, because it is indeed what it is. I think more often of Agnes, age ninety-one, "I don't feel my age, I just feel happy," she said. But the silly sadness is my emotions trying to keep up with my mind. And I think I should let myself feel some sadness, because it is a loss.

Today I have been a whirling dervish. I baked two zucchini breads, a banana bread, an apple crisp, made a double batch of zucchini fritters, and the last two of four loaves of "pain ordinaire" is baking as I write. I washed my car and cleaned up the house.

Now quiet time to write in my journal. I will go for a swim and do some yoga. The yoga is so good for my body and my spirit. Maybe I should stay upside down.

October 5, 2005 *Wonderful dream*
I had this wonderful dream this morning that I had invited all these people for some kind of creative gathering, and everyone loved it, and I was joyful. I felt so alive in my dream, and it was like I was a different person, and the other me was just an empty shell of what I should be. Like gray vs. intense colors. I know that the person I was in my dream is who I really am and should be. I believe the dream was a message to share my writing with others.

October 17, 2005 *Full Harvest Moon*
Low low tide. I took the kayak out with the full moon beaming across the river. Paddled quietly, pausing to listen and watch. Two big splashes, beavers startled me, made me laugh. I heard voices from two fishing boats upriver. Marsh aglow with moonlight. Barred owls whooo whoooing back and forth. And screech owls

whinnying up the forested banks. A line of moonlight
swimming across the water.

October 22, 2005 *Moon of Falling Leaves*
Connor is six months old today. Candles flicker on
my desk. Yellow leaves swirl to the ground. A gray
day. I walked the fields and roads. A box turtle was on
the side of the road and he said, "ssssss," and I picked
him up and said "ssssss" back. On the riverbank,
I found another branch on the ground thick with
coral persimmon, touched its small roundness, its
colour. I swam two mornings ago before work, my
back hurting, my spirit not wanting to give up my
swimming. And yesterday, a long slow kayak ride,
pulling into secret places beneath old trees hanging
over the river, a turtle slipping from log to water in
the marsh, lavender patches of wildflower, trees turned
red. At Mr. Trimmer's pier, I see my beehive, full
moon, and owl still carved into the clay bank.

November 4, 2005 *A heavy sorrow*
Happy Birthday to me.
Lovely, warm autumn day. Sunshine, yellow and red
trees blooming into blue sky. And yes, I went into the
river for my birthday swim. Cold and alive, glowing.
Spunky me, Webster!
Evening ~ a heavy sorrow. The Virginia Supreme

Court sent the matter of the treaty implications back to circuit court in Newport News.

November 5, 2005 *Keeping the sadness at bay*
A message from a friend echoes what I embrace about fighting for this river, "It's not over till it's over, and it's not over. Right will prevail. Good will overcome evil. The sun will shine." All my sadness that I try to keep at bay, but here it is along with the hope.

November 20, 2005 *Wrenched knee*
This afternoon, I was walking briskly across the yard to the pier, when I stepped into a trench that was covered with leaves and wrenched my knee. I slowly scooted on my butt back to the house. Since it was Sunday, Lee and I spent a long evening at Patient First. I am now in a straight knee brace waiting to have an MRI to find out what damage has been done.

November 23, 2005 *Finding the moon*
The dark of midnight. I wake to the waning half moon rising on the river. I smile in the dark and watch the sky, the light of the moon. I get up to pee, and the sky is wild with stars. For a long while I lie awake and watch as the moon rises, rises, and I love her, and I think that no one can have her, no one can

take her, and we are alone in this night, the moon, the blackness, the stars, an owl whoowhooing, geese calling. I think of the men who have given permits to other men to take this river, who tell lies for power and money, who do not care about the Mattaponi people who have honored this river and lived upon her. I think of the men who have never cared. And I watch the moon again and smile again, and I cry because of love, because of the rising moon that no one can take, because of the men who do not love this river. I breathe and wear the sadness, feel my strength. I find the moon again and go back to sleep.

December 18, 2005 *To be bold*
Cool, sunshine, blue skies, white-cloud reflections in the still river. Sitting atop the riverbank. Bleached marsh grasses, bare stalks of trees. Black birds squawking in the woods. Finally, slowly, I made it outside with this wrenched knee. I miss being a part of this world more than I can say. My spirit yearning; I am edgy and sad, trapped and angry. I want to be about, to accomplish work, to tromp through autumn and into winter. I want to be a wild thing. Shadows of birds fly above me. I look at the river and think of swimming, think of kayaking adventures. When I am healed, I want to be new and strong and bold.

2006 ~ Blooming

March 17, 2006 *Michael's twenty-first birthday*
My Michael's twenty-first birthday, a day to celebrate.
I am sitting on the riverbank, the sky blue with
scattered clouds, windy and chilly. I hear the geese.
It has been oh so long since I have come here to sit
and write. My knee has turned into a hell, a trap of
pain from which I have yet to escape. Two surgeries,
excruciating therapy, and more therapy. Yesterday
a day so low. My husband, my family, my friends
have loved and cared with words, flowers, embraces,
deeds, but I want out. I want to be whole. The shad
fishermen are on the water, and I wave to them…

March 23, 2006 *Blooming forth*
Snowed late in the black of night. When I got up at
midnight, the yard was a blanket of white crystals.
First day of spring, we had snow. Web's third year of
leaving this world. And I listened to my body, to my
instincts, and did some gentle floor yoga. I felt small
gain last night with my knee, and today, much more
ease and hope. I think back to the times I followed my
own path, when Warren and I parted especially. Many

people told me to read this book or that book, to get a really good lawyer, but Warren and I chose kindness. To this day, we care for one another and share time with our children. Advice and help can be good, but in the end, what works best for me is to heed my own wisdom.

Yesterday, for the first time in a while, I felt a loving sense from Web and the spirits. Today, I e-mailed the book publisher, Robert Pruett, and said that I want to publish my journal. It is as if I have been *so very stuck and in such despair and pain*, that now, my will and my spirit are sidestepping any doubts, any negative thoughts, maybe sidestepping my mind altogether and *blooming forth*.

May 5, 2006 *Twelve moons*
I made my way slowly to the pier. It is evening. Small black bird flutters above the river, an eagle flew into the marsh. The sky is white clouds speckled with blue. Bull frogs bellow in the marsh. A duck takes flight, wings flapping against the water. Mountain laurel have bloomed.

I have lost my river rhythm, my touch on the earth. Still my knee is unhealed. I have missed autumn, winter, and now we are in spring. I think I will go into summer, maybe another autumn as well. Twelve moons. But I am on the pier.

June 5, 2006 *The owl, again*
Day's end, coming home from work. I drive to the top of the road and park next to the woods and listen to the birds, watch sunlight splashing through the woods, ferns lacy and lime coloured. I want to find my turkey girl and her babies, but I only cast a hello into the woods and do not see her. I drive up the driveway, and as I round the steep curve, wide graceful wings take flight before me. I stop the car, awe stilling my breath… The bird perches in a tree in the passage of spirits—a magnificent reddish brown owl with light spots. This amazing owl sits for a long while, and I sit quietly. It sees me, and we watch one another. It is like one of the owls from my dream that was trapped in the garage, and in my dream, I opened the doors and they flew off free and wild.

At this time, when I am preparing to publish my Chief journal, pausing to be sure I should move forth, this owl comes before me, and I take this as a message to let my words go out. As I sit here amidst the quiet, the green, this sacred place, I feel a peace and timelessness, like a dream of two worlds joining.

October 22, 2006 *I put on my boots.*
Gray day with autumn's trees blazing through. I put on my boots and headed out into the woods, the river's edge, fields, and ravines. I tromped!

October 27, 2006 *Feather on stone*
A year of loss I hope is passing. My knee is getting better, my time outdoors is returning. My book is coming out before too long. I am a feather on stone. The face of love—what does it look like?

Mattaponi Indian chief passes on
© *Indian Country Today* March 31, 2003.
All Rights Reserved
March 31, 2003
by: Bobbie Whitehead / *Indian Country Today*
MATTAPONI INDIAN RESERVATION, Va. - Described as "the strongest connection" between the remaining Powhatan Indian tribes, Daniel Webster Little Eagle Custalow, chief of the Mattaponi Indians, died March 21 in his home.

Custalow, 90, a descendant of Pocahontas and known by both Indian and non-Indian people for his spirituality, served as the tribe's chief for the past 25 years. During this time, Custalow continued the "Mattaponi Treaty of Tribute to the Commonwealth of Virginia," an annual tribute since the treaty's 1646 signing and 1677 ratification.

When the city of Newport News, Va., sought to build a reservoir that would withdraw water from the Mattaponi River and harm the reservation's shad fishery, Custalow spoke out against the plans. This action, some say, broke the tribe's centuries of silence against political actions affecting them.

Other Virginia chiefs say that Custalow was a very loved man by everyone who met him.

"Webster was one of those kinds of leaders that when he spoke, everyone listened," said Chief Ken Adams of the Upper Mattaponi Tribe. "I remember as a child, even though we were separated by distance, all of the Powhatan Indians were very connected. Webster was one of our strongest connections. When I would go to a place and I would hear Webster speak, I would automatically start to listen because he had a powerful voice, and the words that he spoke, you knew that they would have some significance.

"He wasn't just speaking to be heard he was speaking with authority. I remember the authority in his voice and his deep faith. I believe we have lost one of our finest."

Custalow, who operated a trucking service to haul pulpwood, farmed and fished on the Mattaponi River. He also helped provide community service on the reservation and in King William County during the Great Depression, according to the people who knew him.

"I thought very highly of him, and he will definitely be missed," said William Swift Water Miles, Pamunkey Indian Tribe chief and a Powhatan descendant. "People listened to what he had to say and thought very much of what he had to say."

Born Nov. 14, 1912, on the Mattaponi Indian Reservation, Custalow was the youngest of 10 children and resided on the reservation until his death. He was preceded by his wife, Mary White Feather Custalow in 1993 and three grandchildren, Donald Kuhns, Michael Salmons and Darrell Custalow.

Chief Custalow is survived by nine children: five daughters, Eleanor Pocahontas Cannada and husband, Alton; Edith White Feather Kuhns and husband, Raymond; Dolores Little White Dove Salmons and husband, Ralph; Shirley Little Dove McGowan and husband, Olsson; Debra White Dove Perreco and husband, Louis; and four sons, Dr. Linwood Little Bear Custalow and wife, Barbara; Assistant Chief Carl Lone Eagle Custalow; Ryland Little Beaver Custalow and Leon Two Feathers Custalow and wife, Helen. He had 24 grandchildren, and 32 great-grandchildren.

Chief Little Eagle's body was laid to rest on the Mattaponi Indian Reservation March 24.

Appendix

Reservoir Information

In 1990, the City of Newport News in Virginia began seeking authorization to build a reservoir, sixty-five miles away, that would pump up to seventy-five million gallons of water per day from the Mattaponi River, one of the most pristine on the East Coast. Thus began a battle that continues at this writing and has cost Newport News taxpayers at least twenty-five million dollars as of December 2006. Colonel Allan B. Carroll of the Norfolk District Army Corps of Engineers conducted an unprecedented, comprehensive four-year study and recommended denial of the permit in 2001, citing among numerous reasons that the proposed King William Reservoir (KWR), if approved, would represent the largest authorized destruction of wetlands (437 acres) in the history of the Clean Water Act in Virginia and the mid-Atlantic region. Mattaponi Chief, Carl Custalow voices another reason, (one also listed by Colonel Carroll) "If the KWR project happens, it will affect our tribe environmentally, culturally, and economically. It will eventually mean the demise of our reservation and our heritage." Intense political pressure and money by Newport News kept this proposal going on and on. The following newspaper article illustrates the pattern that is typical of how this issue has continued through the years.

In reversal, reservoir project is back on tap
By Scott Harper, The Virginian-Pilot
© December 15, 2006

Richmond— The King William Reservoir proved Thursday why it is becoming the proverbial cat with nine lives.

First, in 2001, the Army Corps of Engineers in Norfolk rejected the $230 million waterworks project, mostly for environmental concerns. That decision was overturned after intense political pressure and a second review by Army Corps brass in New York.

Next, in 2003, the Virginia Marine Resources Commission denied a permit for the project—which would supply drinking water on the Peninsula for the next 40 years—because key fishing grounds in the Mattaponi River would be damaged. That decision also was reversed, after a lawsuit from the sponsor, Newport News Waterworks, and a second special commission hearing.

Finally, on Thursday, the State Water Control Board reversed its own decision from three months ago. The board this time granted a permit extension that sponsors said was necessary to complete environmental studies and other pre-construction activities—work that was delayed because Newport News was too busy fighting for other government permits.

Thursday's reversal, like most things with the King William project, did not come easy—and it may end up in court.

The water board heard testimony for nearly five hours from a standing-room-only crowd of Newport News officials, lawyers, environmentalists, politicians, engineers, scientists, students and American Indians who live near the proposed reservoir site in King William County.

Then there was the question of whether the board should even be considering the matter—or more precisely, reconsidering the matter.

Newport News had asked the board for a rare "reconsideration" of its permit-extension denial in September, a request that critics said was legally fuzzy.

"Newport News once again is engaged in a multi-pronged attack on a decision that doesn't meet its cause," said opponent Robert Richardson of James City County.

Several board members questioned whether allowing a second hearing on the same issue would open the door to other appeals from other aggrieved parties, clogging the board's docket and making its rulings toothless.

In the end, the board voted 6–1 to reconsider. Members then voted 5–2 for a permit extension, good for as long as five years but requiring Newport News to complete its wetlands-compensation plan by 2010.

At one point, lawyers for Newport News and officials from the Virginia Department of Environmental Quality were drafting permit amendments by hand as activists watched.

"I've been involved in wetlands permitting for nearly 20 years," said Ann Jennings, executive director of the *Chesapeake Bay Foundation in Virginia, "and this is a new low. My confidence in the process is totally shattered. I mean, this is crazy."

*Chesapeake Bay Foundation
King William Reservoir

For more than two decades, the City of Newport News and neighboring localities have sought to construct an environmentally destructive reservoir project in King William County. The project would siphon up to 75 million gallons of water a day from the Mattaponi River into a 1,500-acre reservoir in the Pamunkey River watershed. If built, the King William Reservoir (KWR) would destroy more than 430 acres of pristine nontidal wetlands, flood 21 miles of streams, inundate Native American cultural sites, and threaten restoration of American shad.

King William Reservoir at a Glance

CBF has consistently opposed the King William Reservoir because of:

- Impacts to Chesapeake Bay wetlands. The destruction of 437 wetland acres would represent the largest permitted wetland loss in the Mid-Atlantic region since passage of the Clean Water Act in 1972.
- Threats to Virginia's severely depleted American shad population. The reservoir's water intake pipe will be in the Mattaponi River, Virginia's most important shad spawning river.
- Inflated estimates of Newport News' future water needs and the availability of alternative water sources.

Chesapeake Bay Foundation www.cbf.org

Note

The original KWR design, and all the alternatives examined in 1997, were scaled to provide for a 40 million gallons a day (MGD) need. It has since been determined that the actual regional water need is not 40 MGD on the year 2030, but about 16 MGD in the year 2040.

For the latest information on the King William Reservoir

Chesapeake Bay Foundation
http://www.cbf.org

Alliance to Save the Mattaponi River
http://www.savethemattaponi.org

Bibliography

The Case of Eloise Cobell, Blackfeet Indian. http://www.indiantrust.com

Chesapeake Bay Foundation, http://www.cbf.org

Harper, Scott. "In Reversal, Reservoir Project Is Back on Tap." *The Virginian-Pilot,* December 15, 2006.

Nerburn, Kent, and Louise Mengelkoch. *Native American Wisdom*. Novato: CA, New World Library, 1991.

Shnayerson, Michael. "George W. Bush vs. the Environment." *Vanity Fair,* September, 2003.

Testimony of Mark Maryboy, Chairperson Navajo Nation Council Transportation and Community Development Committee. http://financialservices.house.gov/media/pdf/050304mm.pdf

Whitehead, Bobbie. "Mattaponi Indian Chief Passes On." *Indian Country Today*. March 31, 2003.